The Individual and the Welfare State

Axel Börsch-Supan · Martina Brandt ·
Karsten Hank · Mathis Schröder
Editors

The Individual and the Welfare State

Life Histories in Europe

Editors
Prof.Dr. Axel Börsch-Supan
Mannheim Research Institute for
the Economics of Aging (MEA)
L 13, 17
68131 Mannheim
Germany
axel@boersch-supan.de

Prof.Dr. Karsten Hank
University of Cologne
Research Institute for Sociology
Greinstr. 2
50939 Cologne
Germany
hank@wiso.uni-koeln.de

Dr. Martina Brandt
Mannheim Research Institute for
the Economics of Aging (MEA)
L 13, 17
68131 Mannheim
Germany
brandt@mea.uni-mannheim.de

Mathis Schröder, Ph.D.
German Institute for Economic Research
(DIW Berlin)
Mohrenstr. 58
10117 Berlin
Germany
mschroeder@diw.de

This book has been published thanks to the support of the European Commission, Directorate-General for Research, 6th Framework Programme for Research-Socio-economic Sciences and Humanities (contract nr 028812).

The information and views set out in this book are those of the author[s] and do not necessarily reflect the official opinion of the European Communities. Neither the European Communities institutions and bodies nor any person acting on their behalf may be held responsible for the use which may be made of the information contained therein.

The SHARELIFE project is co-funded by the European Community FP6 2002-2006, under the Socio-economic Sciences and Humanities programme.

ISBN 978-3-642-17471-1   e-ISBN 978-3-642-17472-8
DOI 10.1007/978-3-642-17472-8
Springer Heidelberg Dordrecht London New York

Library of Congress Control Number: 2011925540

© Springer-Verlag Berlin Heidelberg 2011
This work is subject to copyright. All rights are reserved, whether the whole or part of the material is concerned, specifically the rights of translation, reprinting, reuse of illustrations, recitation, broadcasting, reproduction on microfilm or in any other way, and storage in data banks. Duplication of this publication or parts thereof is permitted only under the provisions of the German Copyright Law of September 9, 1965, in its current version, and permission for use must always be obtained from Springer. Violations are liable to prosecution under the German Copyright Law.
The use of general descriptive names, registered names, trademarks, etc. in this publication does not imply, even in the absence of a specific statement, that such names are exempt from the relevant protective laws and regulations and therefore free for general use.

Cover design: WMXDesign GmbH, Heidelberg, Germany

Printed on acid-free paper

Springer is part of Springer Science+Business Media (www.springer.com)

List of Authors

Viola Angelini is Assistant Professor at the Department of Economics, Econometrics and Finance of the University of Groningen and Netspar research fellow. She previously worked at the University of Padua as SHARE Post-Doctoral Research Fellow and she holds a joint Ph.D. in Economics from the Universities of York and Padua. Her research fields are applied microeconomics, microeconometrics, household consumption and saving behaviour and the economics of ageing.

Mauricio Avendano works as research fellow at the Department of Public Health of the Erasmus MC in Rotterdam, The Netherlands, where he also obtained his Ph.D. in Epidemiology and Public Health in 2006. His research focuses on international comparisons of health according to socioeconomic position in Europe and the United States. He has recently been awarded several excellence fellowships to study the multiple mechanisms by which social processes influence health in different societies. He has published numerous papers in international journals on this area.

Radim Bohacek is currently a senior researcher at the Economics Institute of the Academy of Sciences. He has been at CERGE-EI since 2000. His research focuses on general equilibrium models with heterogeneous agents, optimal fiscal and monetary government policies, and contract theory. He received his Ph.D. in Economics at the University of Chicago in 1999 and holds M.A. in Economics as well as Public Policy Studies and History from the University of Chicago.

Axel Börsch-Supan is Professor for Macroeconomics and Public Policy and Director of the Mannheim Research Institute for the Economics of Aging at the University of Mannheim, Germany. He holds a Diploma in Mathematics from Bonn University and a Ph.D. in Economics from M.I.T. He started teaching at Harvard's Kennedy School of Government, then taught at Dortmund and Dresden, Germany. Börsch-Supan chairs the Council of Advisors to the German Economics Ministry and is member of the German Academy of Sciences.

Martina Brandt is a Senior Researcher at the Mannheim Research Institute for the Economics of Aging, University of Mannheim, Germany. She is a sociologist, and her current research interests are ageing, social networks and unemployment. She studied at the Universities of Cologne and Zuerich, where she did her doctorate on intergenerational solidarity in Europe. Her dissertation was awarded the German Award for Young Academics ("Deutscher Studienpreis") in 2009.

Agar Brugiavini is Professor of Economics at the University Ca' Foscari of Venice, Italy. She obtained a Ph.D. in Economics at the London School of Economics, UK, and was a lecturer in Finance at the City University Business School, UK. She was a Fulbright Fellow at Northwestern University (USA) and she is currently responsible for the "Employment and Pension" modules of SHARE. She is also a research associate of the Institute for Fiscal Studies, UK and part of the International Social Security ISS project of the National Bureau of Economic Research.

Danilo Cavapozzi is research officer at the University of Padova. He holds a Laurea degree in Statistics and Economics and a Ph.D. in Economics and Management from the same university. His research interests cover microeconometrics, labour economics, the effects of technological change on the labour market and the relationship between health and socioeconomic status.

Dimitris Christelis has a B.A. in Economics from the Athens School of Economics and Business and a Ph.D. in Economics from the University of Pennsylvania. He is currently a SHARE Research Officer at the Universities of Padua and Venice. His research interests include household saving and portfolio investment, microeconometrics, and imputation of missing data.

Sarah Cornaz is a research fellow at the Institute of Social and Preventive Medicine of Lausanne. She holds a Diploma in Geography from Lausanne University.

Caroline Dewilde is Assistant Professor of Sociology at the University of Amsterdam. Her main research interest is the dynamic study of social inequality at different levels of analysis (from the individual life course to the welfare state), with special attention to cross-national variations related to the impact of cultural patterns, family structures and institutional arrangements. Current research projects focus on the cumulative stratification over the life course in different institutional settings and on the long-term interplay between welfare states, housing regimes and life courses.

Loretti Isabella Dobrescu (Ph.D. Economics and Management, University of Padua 2008) is an assistant professor at the Australian School of Business, University of New South Wales, Sydney Australia. She previously held research officer positions at the University of Venice and University of Padua and was a visiting researcher at the Boston University. She was part of the backbone team that developed SHARE, being involved in the "Earnings and Pensions" section and the development of dataset's imputation phase. She is currently affiliated with the Australian Institute for Population Ageing Research (AIPAR) and is active in the research fields of Microeconometrics, with particular emphasis on Economics of Ageing, Health Economics and Mathematics.

Alessio Fiume is SHARE research officer at the Department of Economic Sciences of the University of Padua. He holds a Laurea degree in Statistics for Economics, Business and Finance from the University of Padua.

Christelle Garrouste is a post-doctoral researcher at the Joint Research Center of the European Commission. She holds an MA in International Economics from the

University of Orléans, an MA in International Relations from the University of Auvergne and a Ph.D. in Comparative Education and Economics of Education from the University of Stockholm. Her research interests are in labour economics, education economics and development economics.

Thomas Georgiadis is a Ph.D. student at Panteion University in Athens. His Ph.D. research focuses on poverty analysis. He holds an M.Sc. in Development Economics (University of Sussex, UK). He has been a member of the Greek team in SHARE since 2005.

Karsten Hank was German Survey Manager for Waves 1–3 of SHARE at the Mannheim Research Institute for the Economics of Aging. He is now Professor of Sociology at the University of Cologne and Research Professor at DIW Berlin. His main research interests are in the fields of intergenerational relations, productive ageing, and cross-national analysis. Recent articles have been published in the European Sociological Review, the Journal of Marriage and Family, and The Gerontologist.

Alberto Holly is Professor Emeritus and Visiting Professor at the University of Lausanne, as well as Visiting Professor at the Nova School of Business and Economics, Lisbon, Portugal. He has numerous publications in Econometrics and in Health Economics. He has a wide range of areas of interest in Health Economics such as, the regulation of health care systems, the risk adjustment mechanisms, the diffusion of technological advance in the medical field, and the health economics of ageing.

Julie Korbmacher is a research fellow at the Mannheim Research Institute for the Economics of Ageing where she works for the SHARE database management. She holds a Diploma in Social Sciences from the University of Mannheim and currently works on her Ph.D. thesis, focusing on methodological aspects of the SHARE survey.

Anne Laferrère has been Country Team Leader of SHARE since 2005. She is Senior Research Associate at the Centre de Recherche en Economie et Statistique (CREST) in Paris and her research interests focus on housing policy and family economics. She is Administrateur de l'INSEE (Institut National de la Statistique et des Etudes Economiques) and graduated from ENSAE (Ecole nationale de la Statistique et de l'Administration Economiques). Her life at INSEE before SHARE included that of head of the Housing department, in charge of national housing surveys and housing price indexes.

Philippe Lambert obtained a Master in biostatistics and Ph.D. in statistics from Universiteit Hasselt (Belgium) in 1995. He is professor in statistics at the Université catholique de Louvain (UCL) since 1999 and at the Université de Liège since 2007. He is also member of the Institute of statistics, biostatistics and actuarial sciences (ISBA) at UCL. His research interests are in statistical modelling with specific interest in dependence modelling and in Bayesian smoothing methods.

Karine Lamiraud is an Assistant Professor of Economics at ESSEC Business School in Paris and an Invited Professor at the University of Lausanne. Her current work includes studies of competition and consumer choices in health insurance markets, patient preferences for alternative treatments, health care consumption and costs. Karine has published in Journal of Economic Behavior and Organization, Health Economics, Medical Care and Health Policy. She received the award for "best paper published" for the period 2006–2007 in Health Economics.

Antigone Lyberaki is a Professor of Economics at Panteion University in Athens. She holds a M.Phil. in Development Studies and a Ph.D. in Economics from Sussex University. She acted as coordinator for three Greek National Action Plans for Employment between 2001 and 2003. Her current research interests are ageing societies, gender and migration. She has published extensively in the fields of small and medium-sized firms, employment, migration and gender.

Johan P. Mackenbach is Professor of Public Health at Erasmus MC, University Medical Center Rotterdam, The Netherlands. He is a registered epidemiologist and public health physician, and has published widely in the fields of social epidemiology, medical demography and health services research. He is an elected member of the Royal Dutch Academy of Arts and Sciences, and is actively engaged in exchanges between research and policy.

Karine Moschetti is a project manager in health economics at the Institute of Health Economics and Management, University of Lausanne and University Hospital of Vaud, Lausanne. Her current work includes studies of health care consumption and costs, pharmaceuticals consumption and studies in the field of economic evaluation.

Alberto Motta (Ph.D. Economics and Management, University of Padua 2008) is an assistant professor at the Australian School of Business, University of New South Wales, Sydney Australia. He is affiliated with the Australian Institute for Population Ageing Research (AIPAR) and he is currently working on topics related to the Economics of Ageing, at both microeconomic and macroeconomic level. Other research interests include Microeconomic Theory and Organizational Economics.

Michał Myck is Director of the Centre for Economic Analysis, CenEA, in Szczecin (PL). Since 2005 he has been the Polish Country Team Leader for the Survey of Health, Ageing and Retirement in Europe (SHARE) cooperating with the University of Warsaw and the Mannheim Institute for the Economics of Aging.

Owen O'Donnell is an Associate Professor of Applied Economics at the University of Macedonia and Erasmus University Rotterdam, and during 2009–2010 was a Visiting Professor at the University of Lausanne. He is a Research Fellow of the Tinbergen Institute and NETSPAR, and an Associate Editor of Health Economics. He has published a number of journal articles mainly in the field of health economics and is a co-author of Analyzing Health Equity Using Household Survey Data.

Zeynep Or is a senior economist at the IRDES. Between 1994 and 2003, she worked as an economist and a consultant for the OECD. She has published a number of studies, as well as a book, on the themes of variations in health care use and outcomes across countries and the interaction between institutional and policy settings and health system performance.

Mario Padula is Associate Professor of Econometrics at the University Ca' Foscari of Venice, Italy. He has a Master in Economics from Università Bocconi and a Ph.D. in Economics from University College London. His current research interests are pension reforms, the relation between health and saving, the dynamic properties of expenditures on durable goods and the effect on credit allocation of law enforcement.

Omar Paccagnella is an Assistant Professor of Economic Statistics at the Faculty of Statistics, University of Padua. He graduated in Statistics and Economics in 1998 and obtained a Ph.D. in Applied Statistics in 2003 at the University of Padua. His main research activities currently focus on economics of ageing, anchoring vignettes and multilevel modelling. He has been working in SHARE since the beginning.

George Papadoudis is an economist and researcher and holds a Ph.D. from Panteion University on social inequalities and political economy of ageing. He is the country survey operator of the SHARE project family in Greece (2003–2010). He has worked as a research associate in National Centre of Social Research and as a scientific associate for the Institute of Strategic and Development Studies. He served at the Department of Strategic Planning of the Prime Minister's Political Office and also participated in specialized working groups formed by the National Statistical Service of Greece. His recent work has been focused on the fields of economic and health inequalities, social policy, employment issues and ageing multidimensional patterns.

Giacomo Pasini is a SHARE research officer at the Department of Economics, University Ca' Foscari of Venice, Italy and a Netspar research fellow. He graduated in Economics and Statistics at University of Padua and obtained his Ph.D. in Economics from the University of Venice. His research field is micro-econometrics, in particular the estimation of economic models accounting for social interactions and network effects.

Franco Peracchi is a Professor of Econometrics at "Tor Vergata" University in Rome. He holds a M.Sc. in Econometrics from the London School of Economics and a Ph.D. in Economics from Princeton University. He started teaching at UCLA and NYU, then taught at Udine and Pescara in Italy and Universidad Carlos III in Spain. His research interests include econometric theory and methods, nonparametric and robust statistical method, labour economics, and the economics of social security and pensions.

Sergio Perelman obtained his first degree in Economics at the University of Buenos Aires and a Ph.D. in Economics at the University of Liège, Belgium, where he is professor of economics since 1991. His research interests are social security, labour economics and productivity analysis. Since 2004 he is the Belgium SHARE French speaking Country Team Leader.

Pierre Pestieau received his Ph.D. from Yale. He has taught economics at Cornell University and then at the University of Liege till 2008. He is now Professor Emeritus at the University of Liege. He is also is member of CORE, Louvain-la-Neuve, associate member of PSE, Paris, CEPR Fellow and CESIfo Fellow. His major interests are pension economics, social insurance, inheritance taxation, redistributive policies and tax competition.

Henning Roth is a Ph.D. student in Economics at the University of Mannheim, Germany. He obtained a Diploma in Economics from the University of Mannheim and followed 1 year of his studies at the University of Toronto, Canada.

Brigitte Santos-Eggimann is Professor of Social and Preventive Medicine at the Faculty of Biology and Medicine of the University of Lausanne, Switzerland. She holds a Ph.D. in Medicine from the University of Geneva (Switzerland) and a Ph.D. in Public Health from the Johns Hopkins University, School of Hygiene and Public Health (USA). She is heading the Health Services Unit of the Institute of Social and Preventive Medicine of the University of Lausanne.

Jerome Schoenmaeckers finished his Master Degree in Economics at the University of Liège in 2009. Since then he is the Belgian SHARE French speaking Operator and research assistant at the Department of Economics of the University of Liège.

Mathis Schröder obtained his Ph.D. in Economics from Cornell University in Ithaca, NY, in 2006. He was then project manager of the SHARELIFE project at MEA in Mannheim from 2007 to 2010. He is currently a senior researcher in the German Socioeconomic Panel (SOEP) at the German Institute for Economic Research (DIW) in Berlin, Germany. His research interests include health economics and survey methodology.

Johannes Siegrist is Professor of Medical Sociology and Director of the School of Public Health at the Medical Faculty, University of Duesseldorf, Germany. He received his Ph.D. in Sociology from the University of Freiburg. His major research field is "social determinants of health in midlife and early old age, with a special focus on stressful psychosocial work environment". He recently directed a European Science Foundation program on social variations in health expectancy in Europe. Honors include membership of Academia Europaea.

Nicolas Sirven, Ph.D., is an economist and a Research Fellow at the Institute for research and information in health economics (IRDES). He wrote a Ph.D. on social capital and economic development, and completed a 2 years post-doctorate at the University of Cambridge. He specialised on economic analysis of health status and risk factors at the population level as well as on issues related to ageing of the population.

Jacques Spagnoli is a statistician at the Institute of Social and Preventive Medicine Health Services Research Unit. He holds a Diploma in Mathematics from Lausanne University.

Platon Tinios, is an economist, assistant professor at the University of Piraeus. He studied at the Universities of Cambridge and Oxford. He served as Special Advisor to the Prime Minister of Greece from 1996 to 2004, specializing in the economic analysis of social policy. He was a member of the EU Social Protection Committee from 2000 to 2004. His research interests include ageing populations, social policy, labour economics and public finance. He is the author of research papers and books on pensions and social security reform.

Elisabetta Trevisan is a SHARE Post-doc research fellow at the Department of Economics, University Ca' Foscari of Venice, Italy and a Netspar research fellow. She holds a Laurea degree in Economics and a Ph.D. in Economics from the

University of Venice. Her research interests cover labour economics, economics of aging and evaluation of public policies.

Karel Van den Bosch is a senior researcher at the Herman Deleeck Centre for Social Policy, University of Antwerp. Being educated as a sociologist at the University of Nijmegen in the Netherlands, he holds a doctorate in Political and Social Sciences at the University of Antwerp. His research interests are poverty, income inequality and the impact of welfare states on these. He teaches a course in multivariate analysis.

Aaron Van den Heede is research assistant at the University of Antwerp. He holds a Master's degree in Sociology from the University of Ghent. His main research interests are social stratification, poverty, well-being and ageing.

Morten Wahrendorf is a research fellow at the Department of Medical Sociology of the University Düsseldorf, Germany, where he also obtained his Ph.D. in Sociology in 2009. He holds a Master in Sociology from the University of Montréal, Canada. 2009 he has been awarded a post-doc fellowship from the German Academic Exchange Service to work at the International Centre for Life Course Studies in Society and Health at Imperial College London. His research interests include life course influences on health and well-being in early old age, work stress and retirement behaviour, health inequalities and welfare policy, and survey methodology and statistics.

Guglielmo Weber (Ph.D. economics, LSE 1988) is Professor of Econometrics at the Statistics Faculty of University of Padova and a member of the Economics Department. He previously worked at University College London and University Ca' Foscari of Venice and was a visiting professor at Northwestern University. He is also an international research affiliate of the Institute for Fiscal Studies (London) and CEPR research fellow. He has published papers in several peer-refereed journals, and has been an editor or associate editor of several academic journals in Economics.

Contents

1 **Employment and Health at 50+: An Introduction to a Life History Approach to European Welfare State Interventions** 1
Axel Börsch-Supan and Mathis Schröder

Part I Income, Housing, and Wealth

2 **Explaining Persistent Poverty in SHARE: Does the Past Play a Role?** 19
Platon Tinios, Antigone Lyberaki, and Thomas Georgiadis

3 **Childhood, Schooling and Income Inequality** 31
Danilo Cavapozzi, Christelle Garrouste, and Omar Paccagnella

4 **Human Capital Accumulation and Investment Behaviour** 45
Danilo Cavapozzi, Alessio Fiume, Christelle Garrouste, and Guglielmo Weber

5 **The Impact of Childhood Health and Cognition on Portfolio Choice** 59
Dimitris Christelis, Loretti Dobrescu, and Alberto Motta

6 **Nest Leaving in Europe** 67
Viola Angelini, Anne Laferrère, and Giacomo Pasini

7 **Homeownership in Old Age at the Crossroad Between Personal and National Histories** 81
Viola Angelini, Anne Laferrère, and Guglielmo Weber

8 **Does Downsizing of Housing Equity Alleviate Financial Distress in Old Age?** 93
Viola Angelini, Agar Brugiavini, and Guglielmo Weber

9   Separation: Consequences for Wealth in Later Life 103
    Caroline Dewilde, Karel Van den Bosch, and Aaron Van den Heede

Part II   Work and Retirement

10 Early and Later Life Experiences of Unemployment
    Under Different Welfare Regimes 117
    Martina Brandt and Karsten Hank

11 Labour Mobility and Retirement 125
    Agar Brugiavini, Mario Padula, Giacomo Pasini, and Franco Peracchi

12 Atypical Work Patterns of Women in Europe: What Can
    We Learn From SHARELIFE? 137
    Antigone Lyberaki, Platon Tinios, and George Papadoudis

13 Maternity and Labour Market Outcome: Short
    and Long Term Effects 151
    Agar Brugiavini, Giacomo Pasini, and Elisabetta Trevisan

14 Reproductive History and Retirement: Gender Differences
    and Variations Across Welfare States 161
    Karsten Hank and Julie M. Korbmacher

15 Quality of Work, Health and Early Retirement: European
    Comparisons 169
    Johannes Siegrist and Morten Wahrendorf

16 Working Conditions in Mid-life and Participation
    in Voluntary Work After Labour Market Exit 179
    Morten Wahrendorf and Johannes Siegrist

Part III   Health and Health Care

17 Scar or Blemish? Investigating the Long-Term Impact
    of Involuntary Job Loss on Health 191
    Mathis Schröder

18 Life-Course Health and Labour Market Exit in 13 European
    Countries: Results from SHARELIFE 203
    Mauricio Avendano and Johan P. Mackenbach

19 Work Disability and Health Over the Life Course 215
    Axel Börsch-Supan and Henning Roth

20	**Health Insurance Coverage and Adverse Selection**	225
	Philippe Lambert, Sergio Perelman, Pierre Pestieau, and Jérôme Schoenmaeckers	
21	**Lifetime History of Prevention in European Countries: The Case of Dental Check-Ups**	233
	Brigitte Santos-Eggimann, Sarah Cornaz, and Jacques Spagnoli	
22	**Disparities in Regular Health Care Utilisation in Europe**	241
	Nicolas Sirven and Zeynep Or	
23	**Does Poor Childhood Health Explain Increased Health Care Utilisation and Payments in Middle and Old Age?**	255
	Karine Moschetti, Karine Lamiraud, Owen O'Donnell, and Alberto Holly	

Part IV Persecution

24	**Persecution in Central Europe and Its Consequences on the Lives of SHARE Respondents**	271
	Radim Bohacek and Michał Myck	

Contributors

Viola Angelini Department of Economics, Econometrics and Finance, University of Groningen, Nettelbosje 2, 9747 AE Groningen, The Netherlands, v.angelini@rug.nl

Mauricio Avendano Department of Public Health, Erasmus MC, Room AE-104, P.O. Box 2040, 3000 CA Rotterdam, The Netherlands, mavendan@hsph.harvard.edu

Radim Bohacek The Economics Institute of the Academy of Sciences of the Czech Republic, Politickych veznu 7, 111 21 Prague 1, Czech Republic, radim.bohacek@cerge-ei.cz

Axel Börsch-Supan Mannheim Research Institute for the Economics of Aging (MEA), L13, 17, 68131 Mannheim, Germany, axel@boersch-supan.de

Martina Brandt Mannheim Research Institute for the Economics of Aging (MEA), L13, 17, 68131 Mannheim, Germany, brandt@mea.uni-mannheim.de

Agar Brugiavini Dipartimento di Scienze Economiche, Università Ca' Foscari Venice, Cannaregio, 873, 30121 Venice, Italy, brugiavi@unive.it

Danilo Cavapozzi Dipartimento di Scienze Economiche "Marco Fanno", Università di Padova, Via del Santo 33, 35123 Padova, Italy, danilo.cavapozzi@unipd.it

Dimitris Christelis Department of Economics "Marco Fanno", University of Padua, Via del Santo, 33, 35123 Padua (PD), Italy, dimitris.christelis@gmail.com

Sarah Cornaz Health Services Unit, Institute of Social and Preventive Medicine, University of Lausanne, Route de Berne 52, 1010 Lausanne, Switzerland, sarah.cornaz@chuv.ch

Caroline Dewilde Department of Sociology and Anthropology, University of Amsterdam, OZ Achterburgwal 185, 1012 DK Amsterdam, The Netherlands, c.l.dewilde@uva.nl

Loretti Isabella Dobrescu School of Economics, Australian School of Business, University of New South Wales, Sydney 2052 NSW, Australia, dobrescu@unsw.edu.au

Alessio Fiume Dipartimento di Scienze Economiche "Marco Fanno", Università di Padova, Via del Santo 33, 35123 Padova, Italy, alessio.fiume@unipd.it

Christelle Garrouste European Commission – Joint Research Centre (EC – JRC), Institute for the Protection and Security of the Citizen (IPSC), Unit G.09 Econometrics and Applied Statistics (EAS), TP361 – Via Enrico Fermi 2749, 21027 Ispra (VA), Italy, christelle.garrouste@jrc.ec.europa.eu

Thomas Georgiadis Department of Regional and Economic Development, Panteion University of Social and Political Science, 136, Syngrou Avenue, Athens 17671, Greece, th.georgiadis@gmail.com

Karsten Hank University of Cologne, Institute of Sociology, Greinstr. 2, 50939 Cologne, Germany, hank@wiso.uni-koeln.de

Alberto Holly Institute of Health Economics and Management (IEMS), University of Lausanne, Bâtiment Vidy, Route de Chavannes 31, 1015 Lausanne, Switzerland, Alberto.Holly@unil.ch

Julie Korbmacher Mannheim Research Institute for the Economics of Aging (MEA), L13, 17, 68131 Mannheim, Germany, korbmacher@mea.uni-mannheim.de

Anne Laferrère Insee, 18 boulevard A. Pinard, 75675 Paris, Cedex 14, Italy, anne.laferrere@insee.fr

Philippe Lambert CREPP – HEC Université de Liège, Bd. du Rectorat 7 (B31), 4000 Liege, Belgium, p.lambert@ulg.ac.be

Karine Lamiraud Department of Economics, ESSEC Business School, Avenue Bernard Hirsch, B.P. 50105, 95021 Cergy, France, lamiraud@essec.fr

Antigone Lyberaki Department of Regional and Economic Development, Panteion University of Social and Political Science, 136, Syngrou Avenue, Athens 17671, Greece, antiglib@gmail.com

Johan P. Mackenbach Department of Public Health, Erasmus MC, University Medical Center Rotterdam, Room number Ae-228, P.O. Box 2040, 3000 CA Rotterdam, The Netherlands, j.mackenbach@erasmusmc.nl

Karine Moschetti Institute of Health Economics and Management (IEMS), University of Lausanne, Bâtiment Vidy, Route de Chavannes 31, 1015 Lausanne, Switzerland, Karine.Moschetti@unil.ch

Alberto Motta School of Economics, Australian School of Business, University of New South Wales, Sydney 2052, NSW, Australia, motta@unsw.edu.au

Michał Myck Centre for Economic Analysis, CenEA, ul. Krolowej Korony Polskiej 25, 70-486 Szczecin, Poland, mmyck@cenea.org.pl

Owen O'Donnell Department of Balkan, Slavic and Oriental Studies, University of Macedonia, Egnatia 156, Thessaloniki 54006, Greece, ood@uom.gr

Zeynep Or IRDES, Institute for Research and Information in Health Economics, 10, rue Vauvenargues, 75018 Paris, France, or@irdes.fr

Contributors

Omar Paccagnella Dipartimento di Scienze Statistiche, Università di Padova, Via Cesare Battisti, 241/243, 35121 Padova, Italy, omar.paccagnella@unipd.it

Mario Padula Dipartimento di Scienze Economiche, Università Ca' Foscari Venice, Cannaregio, 873, 30121 Venice, Italy, mpadula@unive.it

George Papadoudis Department of Regional and Economic Development, Panteion University of Social and Political Science, 136, Syngrou Avenue, Athens, 17671 Greece, gpapadoudis@gmail.com

Giacomo Pasini Dipartimento di Scienze Economiche, Università Ca' Foscari Venice, Cannaregio, 873, 30121 Venice, Italy, giacomo.pasini@unive.it

Franco Peracchi Facoltà di Economia, Università Tor Vergata, via Columbia, 2, 00133 Roma, Italy, franco.peracchi@uniroma2.it

Sergio Perelman CREPP – HEC Université de Liège, Bd. du Rectorat 7 (B31), 4000 Liege, Belgium, sergio.perelman@ulg.ac.be

Pierre Pestieau CREPP – HEC Université de Liège, Bd. du Rectorat 7 (B31), 4000 Liege, Belgium, p.pestieau@ulg.ac.be

Henning Roth Mannheim Research Institute for the Economics of Aging (MEA), L13, 17, 68131 Mannheim, Germany, roth@mea.uni-mannheim.de

Brigitte Santos-Eggimann Health Services Unit, Institute of Social and Preventive Medicine, University of Lausanne, Route de Berne 52, 1010 Lausanne, Switzerland, brigitte.santos-eggimann@chuv.ch

Jerome Schoenmaeckers CREPP – HEC Université de Liège, Bd. du Rectorat 7 (B31), 4000 Liege, Belgium, jerome.schoenmaeckers@ulg.ac.be

Mathis Schröder German Institute for Economic Research (DIW), Mohrenstr. 58, 10117 Berlin, Germany, mschroder@diw.de

Johannes Siegrist Department of Medical Sociology, University of Duesseldorf, Universitätsstrasse 1, 40225 Duesseldorf, Germany, siegrist@uni-duesseldorf.de

Nicolas Sirven IRDES, Institute for Research and Information in Health Economics, 10, rue Vauvenargues, 75018 Paris, France, sirven@irdes.fr

Jacques Spagnoli Health Services Unit, Institute of Social and Preventive Medicine, University of Lausanne, Route de Berne 52, 1010 Lausanne, Switzerland, jacques.spagnoli@chuv.ch

Platon Tinios Piraeus University, Karaoli and Dimitriou Street 80, Piraeus 18534 Greece, ptinios@otenet.gr

Elisabetta Trevisan Dipartimento di Scienze Economiche, Università Ca' Foscari Venice, Cannaregio, 873, 30121 Venice, Italy, trevisel@unive.it

Karel Van den Bosch Herman Deleeck Centre for Social Policy, University of Antwerp, Sint Jacobstraat 2, 2000 Antwerp, Belgium, karel.vandenbosch@ua.ac.be

Aaron Van den Heede Herman Deleeck Centre for Social Policy, University of Antwerp, Sint Jacobstraat 2, 2000 Antwerp, Belgium, aaron.vandenheede@ua.ac.be

Morten Wahrendorf Department of Medical Sociology, University of Duesseldorf, Universitätsstrasse 1, 40225 Duesseldorf, Germany, wahrendorf@uni-duesseldorf.de

Guglielmo Weber Dipartimento di Scienze Economiche "Marco Fanno", Università di Padova, Via del Santo 33, 35123 Padova, Italy, guglielmo.weber@unipd.it

Chapter 1
Employment and Health at 50+: An Introduction to a Life History Approach to European Welfare State Interventions

Axel Börsch-Supan and Mathis Schröder

1.1 A New Approach to Analysing the European Welfare State

Health and employment are key determinants of our well-being. They are major objectives of the European welfare state, e.g. of the Lisbon agenda. Yet, health and employment vary tremendously across Europe. This variation is particularly large at older ages when the sum of influences over the entire life course expresses itself.

One example is "healthy life expectancy", a statistic computed by the World Health Organization (WHO 2004) which describes the years from birth to a major disabling health event. It varies by more than 10 years in the European Union. It differs to an astounding extent even across the most highly developed countries. For example, the Swiss enjoy more than three more healthy years of life than residents in Great Britain.

Well known are also the large differences in the share of older individuals who still participate in the labour market. That share, referring to those aged between 55 and 64, varies between 37.2% in Belgium and 74.0% in Sweden (OECD Employment Outlook 2010). Similarly, the share of employed women in all ages varies between 51.1% in Italy and 77.3% in Denmark. That variation has been even greater in the past such that the share of women with their own pensions varies greatly within Europe.

Why are these differences so pronounced? To what extent have these differences been created by policy interventions? The first aim of this book is to shed light on the specific mechanisms through which welfare state interventions may be responsible for these large international differences in health and employment at older

A. Börsch-Supan (✉)
Mannheim Research Institute for the Economics of Aging (MEA), L13, 17, 68131 Mannheim, Germany
e-mail: axel@boersch-supan.de

M. Schröder
German Institute for Economic Research (DIW) Berlin, Mohrenstr. 58, 10117 Berlin, Germany
e-mail: mschroder@diw.de

ages. More ambitiously, a second aim is to translate such findings into improved policy design in the European welfare states.

This is actually not a new topic. Meters of shelf space have been filled with analyses of the welfare state and their policy implications. This book, however, presents an innovative and eye-opening approach to those still most important questions. The common main innovation of the 23 analytical chapters in this book is a combination of life-history micro data with a macro data base of historical welfare state interventions. All chapters are based on the new third wave of one of the most promising cross-national longitudinal data bases currently available, the data of the Survey of Health, Ageing and Retirement in Europe (SHARE; see Börsch-Supan et al. 2005, 2008).

We will first explain why our methodological innovations open new roads to welfare state analysis. The then following section gives an executive summary of each analytical chapter. The final section of this introduction draws our main conclusions.

1.2 Combining Life History Micro Data with Macro Data on Welfare State Interventions

All of us face welfare state interventions at almost every point in the life course. In early childhood, financial support was given to our parents; education laws affected our adolescent lives; during midlife, we may benefit from unemployment compensation and other income support; and once we retire, pension payments determine our income. Throughout the entire life course, health care provision and housing policies shape our daily life. Of course, each of these interventions does not stand alone – an investment in child health care may reduce sickness later in life, increase productivity and thus reduce the need for unemployment insurance take-up. Identifying the effect of these many welfare state interventions, i.e. establishing a causal link between a specific intervention and a specific outcome is therefore a complex enterprise and a methodological challenge.

Our common methodological framework is based on three powerful features:

- First and foremost, we take a *life history approach*, as we believe that the full effect of welfare state interventions can only be assessed over the entire life course and not by comparing concurrent policies and outcomes (e.g., Mayer 2009). Specifically, we have collected life history micro data to identify *intervention points* at which welfare state policies – such as education, income support programmes, work place regulations, health care systems, old-age and disability pension systems – affect women and men at various points in their lives. Some interventions offset, others amplify each other, and they may have cumulative effects over the life course.
- Second, we use a *multidisciplinary approach* that explicitly accounts for the *interactions between health, work conditions and employment*. Analysing health

or employment in isolation ignores the interactions between health care and labour market policies. These interactions are long-term but we believe that they are crucial in creating different health and employment outcomes.
- Third, we base our analyses on *cross-national comparisons*, in particular an innovative combination of life history, cross-sectional micro and *institutional macro data* that take account of general policy differences as well as the large heterogeneity of life circumstances in EU member countries which make similar policies work differently in different life circumstances.

We collected 28,000 individual life histories in 13 European countries: two Nordic countries: Sweden and Denmark; six Central European countries: Netherlands, Belgium, France, Germany, Austria and Switzerland; two Eastern European countries: Poland and the Czech Republic; and three Mediterranean countries: Spain, Italy, and Greece. The data – called "SHARELIFE" data – were collected between October 2008 and August 2009 with computer-aided personal interviews making use of latest technologies, covering the five most important domains of the life course:

- *Children* (e.g., number of children, maternity leave decisions, pregnancies),
- *Partners* (e.g., number of partners, history for each serious relationship),
- *Accommodations* (e.g., place of birth, amenities during childhood, number of moves),
- *Employment* (e.g., number of jobs, job quality, history of work disability), and
- *Health* (e.g., childhood health, current health, health care usage).

An important feature of our life histories is linkage among these domains. For example, we asked when children were born and then linked this to the employment and income situation at the same time. Similarly, we linked changes in health to changes in marital status and changes in accommodation, to name just two examples.

The collected life histories are part of a larger concept: the Survey of Health, Ageing and Retirement in Europe (SHARE). Since 2004, SHARE has collected data on *health* (e.g., self-reported health, physical functioning, cognitive functioning, health behaviour, use of health care facilities), *psychological status* (e.g., psychological health, well-being, life satisfaction, control beliefs), *economic status* (e.g., current work activity, job characteristics, job flexibility, opportunities to work past retirement age, employment history, pension rights, sources and composition of current income, wealth and consumption, housing, education), and the *social support network* (e.g., assistance within families and social networks, transfers of income and assets, volunteer activities).

The combination of all three data collection efforts gives a detailed picture of the status of each individual in 2004, 2006, plus a view across the entire life course in 2008, ranging from career steps, economic conditions, family history, health development, and housing back to childhood living conditions. The data thus provide a fascinating account of the life in Europe over the past century – a century not only characterized by wars and oppression but also dramatic changes in the extension and influence of the welfare state on individuals' lives.

Collecting retrospective life histories is not easy since memory fails as we all know. We were helped by neuro-psychologists and survey methodologists in developing a sophisticated electronic questionnaire underlying the face-to-face interview to make recollection easier. Each domain was depicted as a graphical time line. The respondents could jump between these time lines, thereby linking events that are easier to remember (such as the birth of a child) with events that are harder to remember (such as a spell of unemployment). The reader is referred to the detailed descriptions in the supplementary volume on the SHARELIFE methodology (Schröder 2011).

One may still be sceptical about the quality of such retrospective data. Recent studies, however, such as Smith (2009) and Haas and Bishop (2010) have validated retrospective data with objective records. Their results suggest that, while caution is clearly warranted and has been applied to the chapters in this book, there is valuable information in retrospective measures that supports a judicious use in population-based research.

In parallel, we have built up a *contextual data base* that describes how welfare state interventions have changed over time and across countries. Typical interventions are education (e.g., years spent in school), medical care (e.g., vaccinations, density of doctors), income support (e.g., unemployment insurance, maternity benefits), pensions (characterized by, e.g., generosity of public and occupational pensions, as well as early retirement age), and work place characteristics (e.g., regulations on work place safety). This information was drawn from a multitude of sources at three levels. First, we used existing synopses at the European level (e.g., the MISSOC data on social services: European Communities 2008). Second, we spliced information from national sources together (e.g., the characterization of educational systems by Garrouste 2010). Finally, we exploited our own SHARE data to create variables describing the environment in which people have lived and worked (e.g., features of the work place).

In addition to a common data base, the analyses in this book are also guided by a common theoretical framework linking welfare state interventions to health and employment, taking – as an important innovation – interactions between health and employment into account which is now possible at the micro level due to the very detailed SHARE data (Fig. 1.1).

Some welfare state interventions affect health and employment *directly*. Early retirement, for example, is directly and often immediately influenced by the rules of the pension, disability and unemployment systems. Health is directly affected by the health care systems. In addition, there are long-run interventions of the welfare state – such as education, preventive health care and work place regulations – which have complex *indirect and interrelated* effects over the life course on both health and employment. Preventive health care, for instance, not only increases health but also employment at older ages. High work place standards do not only improve employment at older ages by reducing early retirement, they also tend to enhance physical and mental health.

Finally, welfare state interventions may have accumulative effects over the life-course as each intervention builds upon earlier interventions. Understanding the *accumulation of direct and indirect* welfare state interventions and their

1 Employment and Health at 50+: An Introduction to a Life History Approach

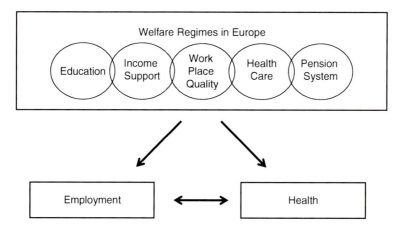

Fig. 1.1 General framework

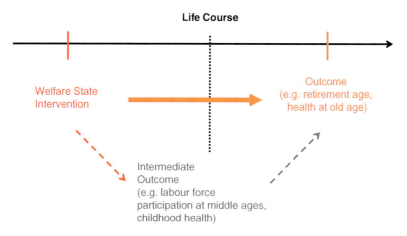

Fig. 1.2 Paradigm 1

interactions over the life cycle in shaping health and employment outcomes is a recurring theme in this volume.

More precisely, the chapters in this book used two common paradigms, depending on the specific mechanism under research.

In paradigm 1, we observe some early or mid-life life welfare state intervention, e.g., education or access to health care, and relate it to later-life outcomes, especially health and/or employment at older ages. In order to isolate the effects of welfare state interventions, we need to carefully correct for other influences over the life course, reaching from childhood health over labour force participation at middle ages to marital status at the time of interview (Fig. 1.2).

In paradigm 2, welfare state interventions modify the influence of early life conditions on later-life outcomes. For example, childhood socio-economic status

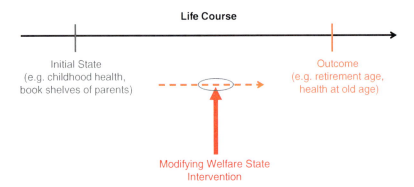

Fig. 1.3 Paradigm 2

is well known to affect later-life health, but the healthcare system in a country may moderate or amplify this inter-temporal link. One benevolent hypothesis to be tested is whether the inter-temporal link is weakened through health-related welfare state interventions by giving extra attention to individuals with a background of childhood poverty. A more cynical hypothesis would claim that state-provided healthcare services will amplify this link because the wealthy have better access to them (Fig. 1.3).

The 23 analyses presented in this book are examples of what the SHARE life history data can uncover, using the above framework. They are kept short und succinct, and are targeted to an audience who wants to see without too much analytical effort how the multitude of welfare state interventions are reflected in the SHARE life histories, and which policy conclusions suggest themselves. We hope that they ignite many more research papers to come.

Some caveats should be kept in mind. First, the analyses in this book are based on the very first data released in spring 2010. They are preliminary in so far as later data releases may correct data errors. Second, since the chapters are kept short and succinct, they cannot employ the full apparatus of modern theory and econometrics, and we therefore apply great care to distinguish associations from causality. Firm policy conclusions, of course, can only be based on the latter. For example, our respondents have survived until the interview. This may create a "survivor bias" since these respondents may be healthier and have been living under luckier circumstances than those who have passed away earlier.

1.3 How the Welfare State Has Shaped Health, Employment and Many Other Aspects of Our Lives at Older Ages

The 23 welfare state analyses are structured by four broad themes. They look at outcomes in later life such as income, housing, wealth, retirement age, volunteering activities, and health status and how these outcomes have been affected by welfare state interventions such as public spending on social protection – from poverty relief over housing subsidies to maternity support – as well as education policies and access to health care over the life course. Finally, we investigated the long-term effects of a very sad chapter of "interventions" in the early lives of many Europeans, namely persecution, especially during the Nazi and Communist regimes.

1.3.1 Part I: Income, Housing, and Wealth

Poverty is one of the most dreaded events in an individual's life, and many political ways have been thought of to reduce poverty. Especially poverty at older ages is problematic, as it relates to poorer health, fewer social contacts and bad economic conditions. *Platon Tinios*, *Antigone Lyberaki* and *Thomas Georgiadis* (Chap. 2) look at how childhood deprivation translates into later life poverty. While they conclude that there are persisting effects, i.e. those deprived early on continue to have a higher risk of poverty in old age, they also find that these effects are soothed by welfare state interventions such as *spending on social protection.*

The transition across different socio economic groups is one of the concerns of policy makers. Do children not only inherit their parents' genes, but also their social status? *Danilo Cavapozzi*, *Christelle Garrouste* and *Omar Paccagnella* (Chap. 3) explore this question by relating the parental financial and educational background with educational attainment and income inequality later in life. They conclude that *education policies* fostering access to education and increasing the number of years spent in full time education might qualify as a possible strategy to reduce income dispersions.

The demographic changes make government pension income more volatile and insecure. For this reason, more and more countries aim at increasing the private pillars of old age provision. But how can people be brought into investing, and how is the investment decision determined by differences in the population? This question is tackled by *Danilo Cavapozzi, Alessio Fiume, Christelle Garrouste* and *Guglielmo Weber* (Chap. 4). They document that age of entry into financial markets varies widely across European countries, with some of these differences relating to income, gender, demographics, and family background. A large part of the variation is also due to human capital accumulation, particularly the accumulation of mathematical skills. The remaining fraction is likely due to institutional differences in access to financial markets. An important policy implication of their analysis is that promoting *better education in mathematics* is likely to reduce differences in access to financial instruments.

Dimitris Christelis, *Loretti Dobrescu* and *Alberto Motta* (Chap. 5) investigate in a similar direction: their paper is concerned with how childhood health and cognition relates to the individual's portfolio choice later in life. Not only is bad childhood health an indicator for fewer investments, the same holds for lack of a usual source of health care during childhood. In addition, performance during school relates positively to the amount of investments, which leads the authors to stress the necessity of welfare policies to intervene early in life to increase access to health care and improve educational attainment.

Housing market interventions are ubiquitous in Europe. They include social housing supply, rent control, and tax subsidies for homeownership, to name only the most prominent ones. Many housing policies are geared specifically to the older population. What are the side effects to the younger generation in an aging society? *Viola Angelini*, *Anne Laferrère* and *Giacomo Pasini* (Chap. 6) investigate the "nest leaving behaviour" across Europe, i.e., the age at which individuals leave their parental home. While it is well-known that Mediterranean children stay longer with their parents than Scandinavians, less is known about how that is affected by housing policies. Angelini, Laferrère and Pasini find that *rent control and tax subsidies for homeownership* increase the nest leaving age, while the *provision of social housing* lowers it.

Besides insuring housing consumption, a home may be seen as a secure asset in case of need. It is also a family asset that may be transmitted to the next generation. *Viola Angelini*, *Anne Laferrère* and *Guglielmo Weber* (Chap. 7) document in their paper the changes in home-ownership rate and age patterns across the European countries and cohorts. They related these changes to *housing policies and credit market development* and show a clear positive relationship between a well developed credit market and the stock of home owners in a country. In addition, the better the housing market, the less likely people are to transfer their housing assets to their kin.

What happens when people are house-rich but cash-poor? This issue is looked at by *Viola Angelini*, *Agar Brugiavini* and *Guglielmo Weber* (Chap. 8), using price data on home purchases and sales. They argue that the importance of trading down as a form of equity release depends on *financial and mortgage markets access*, as well as on the availability of public housing and long term care accommodation. In most European countries financial instruments that allow equity release are unavailable, and cheap public housing is scarce, so trading down is the only way to generate a cash flow out of the available home equity. They show that in those countries where mortgage markets are less well developed, lower fractions of home-owners trade down by selling and buying, and higher fractions of home-owners report financial distress, suggesting that to avoid illiquidity for the elderly across Europe, developing mortgage markets is imperative.

Wealth in later life is typically negatively influenced by family events such as divorce. This effect depends on *divorce laws* which differ across Europe. *Caroline Dewilde*, *Karel Van den Bosch* and *Aaron Van den Heede* (Chap. 9) investigate how marital separation influences later life wealth, a question particularly important as divorce rates are increasing all over Europe in recent years. They find a negative effect over all European countries for divorced women who have remained single. This can be taken as evidence that although women have become more independent through the years, especially the group of today's elderly women are vulnerable to separation from the husband they economically depend on.

1.3.2 Part II: Work and Retirement

Martina Brandt and *Karsten Hank* (Chap. 10) investigate the so-called "scarring effects" of early unemployment on later life employment opportunities. Welfare state interventions, one would hope, should minimize these effects in order to prevent downwards spirals. Their analysis provides clear evidence for scarring effects even among older workers, though. Differences in individuals' unemployment risks across welfare states suggest that *labour market institutions and educational systems* bear in them the potential for significant (positive and negative) interventions affecting people's employment opportunities across the entire life course.

Active labour market policies aim to keep unemployment spells short and labour mobility high, in order to maximize earning capabilities over the life course. However, retirement outcomes are open to debate: one position states that the policies should be reflected in higher retirement income replacement rates, while the other side argues that the Anglo-Saxon model of high job mobility creates low paid jobs and thus lowers pension income. *Agar Brugiavini, Mario Padula, Giacomo Pasini* and *Franco Peracchi* (Chap. 11) shed light on this debate. They do not find a direct translation of job mobility into cross-country differences in retirement income, provided it is tempered by policies that limit long-term unemployment.

Antigone Lyberaki, Platon Tinios and *Georgios Papadoudis* (Chap. 12) document the complexity of women's employment patterns and how these have been shaped by the interaction of individual family experiences with specific welfare state institutions, such as *employment protection and maternity leave regulations*. In younger cohorts, and almost everywhere in Europe, more women exhibit adaptive careers, leaving and re-entering the labour market. The family-work patterns, which used to follow very polarized patterns, thus have somewhat converged, and welfare state policies are shown to have played an important role in this still ongoing development.

One aim of maternity leave provisions is to make sure that maternity does not precipitate a permanent exit from the labour force. Does this welfare state intervention achieve this aim? *Agar Brugiavini, Giacomo Pasini* and *Elisabetta Trevisan* (Chap. 13) compare the labour market participation of women in countries with different *maternity leave provisions*. They then evaluate the resulting retirement income replacement rate which can be interpreted as a measure of life-time earnings. The results by Brugiavini, Pasini and Trevisan are sobering in so far, as countries with generous maternity leave provisions have higher exit rates and lower retirement income replacement rates.

The implications of childbearing history for individuals' labour force participation in later life are not well-investigated yet. *Karsten Hank* and *Julie Korbmacher* (Chap. 14) investigate how men's and women's entry into retirement is associated with parental status and whether this varies across *welfare regimes (with different employment opportunities and pension entitlements for parents)*. They find that mothers are more likely than childless women to exit the labour force early, whereas fathers tend to retire later than other men. The association between childbearing and earlier retirement appears to be particularly strong among women living under a social-democratic or post-communist welfare state regime, that is, in countries exhibiting relatively high levels of female labour force participation.

Avoiding early exits from the labour force is an important policy goal in the European Union. The association between quality of work, health, and early retirement is investigated by *Johannes Siegrist* and *Morten Wahrendorf* (Chap. 15), who also look into the potential role of labour market and social policies in mediating this association. A main finding is that quality of work was generally higher in countries with pronounced *active labour market policies, such as training programmes for older adults*. Similarly, continued employment into old age was more prevalent in countries with high expenditures *in rehabilitation services*.

Older people may contribute to society in productive ways even after retirement, e.g. as volunteers. *Morten Wahrendorf* and *Johannes Siegrist* (Chap. 16) show that elders' propensity to serve as a volunteer today is negatively associated with poor mid-life working conditions, stressing the need to take a life course perspective. Moreover, the authors find the extent of volunteering in early old age to be influenced positively by *policy measures aimed to improve the quality of work and employment*, the extent of *lifelong learning* and the amount of resources spent on *rehabilitation services*.

1.3.3 Part III: Health and Health Care

Does unemployment cause bad health or does bad health lead to unemployment? *Mathis Schröder* (Chap. 17) investigates the association between unemployment and long-term effects on health, using information on business closures to have a causal relationship between unemployment and health. He finds negative health effects of unemployment even up to 40 years after the business closure. In an additional analysis he explores whether the welfare state can mitigate some of the effects of unemployment on health, and finds that especially for women, there are strong positive effects of *unemployment benefits* on long-term health.

In most European countries, long-term illness is associated with earlier exit from the labour market. This well known – but can higher public health investments ameliorate this association? This is the key question posed by *Mauricio Avendano* and *Johan Mackenbach* (Chap. 18). Their results do not generally suggest a strong correlation of the level of public health investments with the prevalence of long-term illness. However, they find that investments in curative health care are strongly (and negatively!) associated with the prevalence of long-term illness. They also find that larger investments in unemployment benefit programmes are associated with a larger impact of illness on labour force participation, suggesting that higher unemployment benefits may potentially work as incentive towards earlier exit from the labour market due to illness.

Disability insurance is an important part of the European welfare state. It insures individuals who are unable to work due to physical or mental health problems at relatively early stages in life against falling into poverty. Striking, however, is the huge variation of individuals receiving disability insurance across Europe. *Axel Börsch-Supan* and *Henning Roth* (Chap. 19) exploit the health histories in the SHARELIFE data to understand whether these international differences are due to bad health at childhood and/or long-term health problems during adult life. While life-course health problems do indeed increase the odds of receiving a disability pension within each country, they do not explain the large international variation. Börsch-Supan and Roth explain this variation with differences in the generosity of the *national disability insurance programmes*.

Adverse selection is still one of the largest problems in the health care markets all over the world. *Philippe Lambert, Sergio Perelman, Pierre Pestieau* and *Jérôme Schoenmaeckers* (Chap. 20) investigate if there is a relation between health risk and insurance coverage, thereby uncovering possible adverse selection. They measure health risks through life course variables such as childhood health and long lifetime illnesses. They relate this to *the health insurance coverage*, but find little evidence of adverse selection across the SHARELIFE countries. Although this may be suggestive to not take policy actions, they argue that further work is needed to actually claim that.

Health as an adult is always related to health care and accessibility of health care throughout one's life. Specifically *dental care* is an important aspect of our daily life, which has changed considerably over the last 50 years. *Brigitte Santos-Eggimann, Sarah Cornaz* and *Jacques Spagnoli* (Chap. 21) take into account the density of dentists when investigating how much dental care older Europeans have received throughout their lives. They report a clear cohort effect – older Europeans suffered more from undercoverage of dental care, although rates are decreasing over the life span. The direct policy implications seem to already be in place – a higher density of dentists will lead to better use of care and improve well-being in later life.

Nicolas Sirven and *Zeynep Or* (Chap. 22) take a more general view on the problem by looking a wide array of *preventive health care measures*, e.g. *blood pressure tests, vision tests, or mammograms*. They report a shift toward more regular care among all countries, however, differences remain between countries and social classes: the higher the education, for example, the higher the propensity to engage in preventive care. Relating the tests to density of doctors, they obtain a similar result as the previous paper: the more the better. Given the dispersion of medical expertise in Europe, these results suggest that there is significant room for public health policies for reducing disparities in regular use of health services within and across European countries.

The question of how childhood conditions affect later life does not only apply to education or social class, but also – and maybe even more so – to health. In the current light of increasing health care costs across the world, this may be especially important. In their paper, *Karine Moschetti, Karine Lamiraud, Owen O'Donnell* and *Alberto Holly* (Chap. 23) show that poor health, parental smoking and limited *access to health care during childhood* are associated with greater utilisation of, and payments for, health care in middle and old age. Interestingly, the association operates mainly through reduced health in adulthood, and less through socioeconomic status. The results are suggestive for policy: improving childhood health in populations now will lead to future cohorts costing less in old age than do their current counterparts.

1.3.4 Part IV: Persecution

The SHARE generation in Europe has experienced many major historical events – among them World War II and the rise and fall of the Communist regimes. However, the population affected by these events is rapidly shrinking, as age takes its toll. In this sense, *Radim Bohacek* and *Michał Myck* (Chap. 24) provide us with a unique analysis: they look at the consequences of *persecution* on people's life, especially on their health and employment careers. They find – even now – strong effects for those who have suffered from persecution and come to the conclusion that while thankfully, in today's Europe, persecution is absent, all the more effort needs to be taken to protect those in other countries suffering from it.

1.4 Conclusions

The SHARE life histories have provided a fascinating account of the life in Europe over the past century. While this century was characterized by wars and oppression, as the last chapter has shown, it has also generated dramatic changes in the extension and influence of the welfare state on individuals' lives. They have, arguably, improved our lives tremendously, and this is reflected in our life histories. First and foremost, health has become much better and life expectancy increased to an extent unprecedented in history. Education has vastly improved. Employment patterns have changed with an enormous increase in female labour force participation and generally later entries into the labour force combined with earlier exits.

The main challenge for the 23 analyses in this book was therefore to isolate specific effects generated by the welfare state in an environment in which many life circumstances dramatically changed. Many of these analyses were indeed able to identify significant and quantitatively important effects of welfare state interventions on later-life outcomes. Education policies, e.g., achieve quite clearly higher retirement incomes and better health in old age. We also find some evidence that long-term policies such as health prevention and life-long learning had positive effects on activity levels and health in old age.

Other analyses find, also quite strikingly, no or ambiguous effects. One example are active labour market policies which do not seem to have influenced labour mobility to an extent that results in higher life-time incomes. Another example were maternity benefits which apparently have reduced rather than increased life-time income and thus resulted in lower public pension benefits to women who have received maternity benefits compared to other women with children. Further research will have to sharpen the analysis until final conclusions can be drawn; in particular, they have to investigate potential counteracting mechanisms.

Some of these findings will be controversial. Some are certainly preliminary and require the apparatus of a more refined statistical methodology. Hopefully, the 23 analyses will inspire our readers to follow up our work with their own analyses by using the SHARE data, especially the newly collected life histories.

Acknowledgements Thanks belong first and foremost to the participants of this study. None of the work presented here and in the future would have been possible without their support, time, and patience. It is their answers which allow us to sketch solutions to some of the most daunting problems of ageing societies. The editors and researchers of this book are aware that the trust given by our respondents entails the responsibility to use the data with the utmost care and scrutiny.

Collecting these data has been possible through a sequence of contracts by the European Commission and the U.S. National Institute on Aging, as well as support by many of the member states.

The SHARE data collection has been primarily funded by the European Commission through the 5th framework programme (project QLK6-CT-2001-00360 in the thematic programme Quality of Life). Further support by the European Commission through the 6th framework programme (projects SHARE-I3, RII-CT-2006-062193, as an Integrated Infrastructure Initiative, COMPARE, CIT5-CT-2005-028857, as a project in Priority 7, Citizens and Governance in a Knowledge Based Society, and SHARE-LIFE (No 028812 CIT4)) and through the 7th framework programme (SHARE-PREP (No 211909) and SHARE-LEAP (No 227822)) is gratefully acknowledged. We thank, in alphabetical order, Giulia Amaducci, Kevin McCarthy, Hervé Pero, Ian Perry, Robert-Jan Smits, Dominik Sobczak and Maria Theofilatou in DG Research for their continuing support of SHARE. We are also grateful for the support by DG Employment, Social Affairs, and Equal Opportunities through Georg Fischer, Ruth Paserman, Fritz von Nordheim, and Jérôme Vignon, and by DG Economic and Financial Affairs through Declan Costello, Bartosz Pzrywara and Klaus Regling.

Substantial co-funding for add-ons such as the intensive training programme for SHARE interviewers came from the US National Institute on Ageing (U01 AG09740-13S2, P01 AG005842, P01 AG08291, P30 AG12815, R21 AG025169, Y1-AG-4553-01, IAG BSR06-11 and OGHA 04-064). We thank John Phillips and Richard Suzman for their enduring support and intellectual input.

Some SHARE countries had national co-funding which was important to carry out the study. Sweden was supported by the Swedish Social Insurance Agency and Spain acknowledges gratefully the support from Instituto Nacional de Estadistica and IMSERSO. Austria (through the Austrian Science Foundation, FWF) and Belgium (through the Belgian Science Policy Administration and the Flemish agency for Innovation by Science and Technology) were mainly nationally funded. Switzerland received additional funding from the University of Lausanne, the Département Universitaire de Médecine et Santé Communautaires (DUMSC) and HEC Lausanne (Faculté des Hautes Etudes Commerciales). Data collection for wave 1 was nationally funded in France through the Caisse Nationale d'Assurance Maladie, Caisse Nationale d'Assurance Vieillesse, Conseil d'Orientation des Retraites, Direction de la Recherche, des Etudes, de l'Evaluation et des Statistiques du ministère de la santé, Direction de l'Animation de la Recherche, des Etudes et des Statistiques du ministère du Travail, Caisse des Dépôts et Consignations, and Commissariat Général du Plan. INSEE (Institut National de la Statistique et des Etudes Economiques) co-founded all 3 waves.

SHARELIFE was a different type of survey than the previous two rounds of interviews, requiring new technologies to be developed and used. Programming and software development for the SHARELIFE survey was done by CentERdata at Tilburg. We want to thank Alerk Amin, Maarten Brouwer, Marcel Das, Maurice Martens, Corrie Vis, Bas Weerman, Erwin Werkers, and Arnaud Wijnant for their support, patience and time. Kirsten Alcser, Grant Benson, and Heidi

Guyer at the Survey Research Center (SRC) of the University of Michigan Ann Arbor again provided the Train-the-Trainer programme for SHARELIFE, and invested tremendous amounts of time and work to develop the prototype of a quality profile for the data collection, which included visiting the sites of the national interviewer trainings in participating countries. Kate Cox, Elisabeth Hacker, and Carli Lessof from the National Centre for Social Research (NatCen) gave helpful input in designing the questionnaire and pointed out the retrospective specifics in the interview process. We always kept in close contact with the professional survey agencies – IFES (AT), PSBH, Univ. de Liège (BE), Link (CH), SC&C (CZ), Infas (DE), SFI Survey (DK), Demoscopia (ES), INSEE (FR), KAPA Research (GR), DOXA (IT), TNS NIPO (NL), TNS OBOP (PL), and Intervjubolaget (SE) – and thank their representatives for a fruitful cooperation. Especially the work of the more than 1,000 interviewers across Europe was essential to this project.

The innovations of SHARE rest on many shoulders. The combination of an interdisciplinary focus and a longitudinal approach has made the English Longitudinal Survey on Ageing (ELSA) and the US Health and Retirement Study (HRS) our main role models. Input into the concepts of retrospective questionnaires came from Robert Belli and David Blane. The life history questionnaire has been implemented first in the ELSA study, and without the help of people involved there (James Banks, Carli Lessof, Michael Marmot and James Nazroo), SHARELIFE could not have been created in such a short time. SHARELIFE has also greatly profited from detailed advice given by Michael Hurd, Jim Smith, David Weir and Bob Willis from the HRS as well as by the members of the SHARE scientific monitoring board: Orazio Attanasio, Lisa Berkman, Nicholas Christakis, Mick Couper, Michael Hurd, Daniel McFadden, Norbert Schwarz and Andrew Steptoe, chaired by Arie Kapteyn. Without their intellectual and practical advice, and their continuing encouragement and support, SHARE would not be where it is now.

Since SHARELIFE was an entirely newly designed questionnaire, the work of developing and constructing the questions was immense. We are very grateful to the contributions of the eight working groups involved in this process. Specifically, Agar Brugiavini, Lisa Calligaro, Enrica Croda, Giacomo Pasini, and Elisabetta Trevisan developed the module for *financial incentives of pension systems*. Johannes Siegrist and Morten Wahrendorf provided input for the module on *quality of work and retirement*. The development of questions for the part of *disability insurance and labour force participation of older workers* was responsibility of Hendrik Jürges, whereas the *health and retirement* section was constructed by Johan Mackenbach and Mauricio Avendano. *Preventive care, health services utilisation, and retirement* fell into the realm of Brigitte Santos-Eggimann and Sarah Cornaz, and Karsten Hank provided his input for the *gender, family, and retirement* section. *Wealth and retirement* questions were designed by Guglielmo Weber and Omar Paccagnella, and finally, questions on *health risk, health insurance, and saving for retirement* were integrated by Tullio Japelli and Dimitri Christelis.

A large enterprise with 150 researchers in 13 countries entails also a large amount of day-to-day work, which is easily understated. We would like to thank Kathrin Axt, Maria Dauer, Marie-Louise Kemperman, Tatjana Schäffner, and Eva Schneider at the MEA in Mannheim for their administrative support throughout various phases of the project. Annelies Blom, Martina Brandt, Karsten Hank, Hendrik Jürges, Dörte Naumann, and Mathis Schröder provided the backbone work in coordinating, developing, and organizing the SHARELIFE project. Preparing the data files for the fieldwork, monitoring the survey agencies, testing the data for errors and consistency are all tasks which are essential to this project. A small glimpse into the details and efforts of data preparation is provided in the methodology volume to this project (Schröder 2010). The authors and editors are grateful to Christian Hunkler, Thorsten Kneip, Julie Korbmacher, Barbara Schaan, Stephanie Stuck, and Sabrina Zuber for data cleaning and monitoring services at the MEA in Mannheim, and Guiseppe de Luca and Dimitri Christelis for weight calculations and imputations in Padua, Salerno and Venice. Finally, Theresa Mutter at the MEA provided excellent assistance in proof-reading the finalized versions of these papers.

Last but by no means least, the country teams are the flesh to the body of SHARE and provided invaluable support: Rudolf Winter-Ebmer, Nicole Halmdienst, Michael Radhuber and Mario Schnalzenberger (Austria); Karel van den Bosch, Sergio Perelman, Claire Maréchal, Laurant

Nisen, Jerome Schoenemaeckers, Greet Sleurs and Aaron van den Heede (Belgium); Radim Bohacek, Michal Kejak and Jan Kroupa (Czech Republic); Karen Andersen Ranberg, Henriette Engberg, and Mikael Thingaard (Denmark); Anne Laferrère, Nicolas Briant, Pascal Godefroy, Marie-Camille Lenormand and Nicolas Sirven (France); Axel Börsch-Supan and Karsten Hank (Germany); Antigone Lyberiaki, Platon Tinios, Thomas Georgiadis and George Papadoudis (Greece); Guglielmo Weber, Danilo Cavapozzi, Loretti Dobrescu, Christelle Garrouste and Omar Paccagnella (Italy); Frank van der Duyn Shouten, Arthur van Soest, Manon de Groot, Adriaan Kalwij and Irina Suanet (Netherlands); Michał Myck, Malgorzata Kalbardczyk and Anna Nicinska (Poland); Pedro Mira and Laura Crespo (Spain); Kristian Bolin and Thomas Eriksson (Sweden); Alberto Holly, Karine Moschetti, Pascal Paschoud and Boris Wernli (Switzerland).

References

Börsch-Supan, A., Brugiavini, A., Jürges, H., Mackenbach, J., Siegrist, J., & Weber, G. (Eds.). (2005). *Health, ageing and retirement in Europe – First results from the survey of health, ageing and retirement in Europe.* Mannheim: MEA.

Börsch-Supan, A., Brugiavini, A., Jürges, H., Kapteyn, A., Mackenbach, J., Siegrist, J., et al. (Eds.). (2008). *Health, ageing and retirement in Europe – Starting the longitudinal dimension (2004–2007).* Mannheim: MEA.

European Communities. (2008). *Mutual Information System on Social Protection (MISSOC).* Brussels: European Communities.

Garrouste, C. (2010). *100 Years of educational reforms in Europe: A contextual database. EUR 24487 EN.* Luxembourg: Publications Office of the European Union.

Haas, S. A., & Bishop, N. J. (2010). What do retrospective subjective reports of childhood health capture? Evidence from the Wisconsin Longitudinal Study. *Research on Aging, 32*(6), 698–714.

Mayer, K. U. (2009). New directions in life course research. *Annual Review of Sociology, 35*, 413–433.

OECD. (2010). *OECD employment outlook.* Paris: OECD.

Smith, J. P. (2009). Reconstructing childhood health histories. *Demography, 46*(2), 387–404.

Schröder, M. (ed.). (2011). *Retrospective Data Collection in the Survey of Health, Ageing and Retirement in Europe. SHARELIFE Methodology.* MEA, Mannheim.

WHO. (2004). *World Health Report 2004.* Geneva: WHO.

Part I
Income, Housing, and Wealth

Chapter 2
Explaining Persistent Poverty in SHARE: Does the Past Play a Role?

Platon Tinios, Antigone Lyberaki, and Thomas Georgiadis

2.1 Identifying 50 Years of Social Progress

Poverty alleviation is certainly the most emblematic of European Union ambitions in the field of social policy – encompassing in a visible and politically salient way the cumulative end effect of many separate interventions in social and economic policy. The question posed in this paper is, therefore: Does the past play a role in the 50+ poverty we see today?

This investigation will proceed in three steps: First, poverty in the two waves of SHARE will be used as the starting point of the analysis. Poverty alleviation is a key motivation for social policy; in a way poverty in our group of 50+ should portray the accumulation of public policies over the life histories of the persons concerned. Old age protection is the most venerable of the objectives of the Welfare State in Europe; the amelioration of the effects of social and economic shocks so that they have no long term effects is a key objective of social policy. Second, a picture of relative deprivation at age 10 is collated from SHARELIFE information and an idea gleaned of its link with poverty status in later life. Childhood poverty is a key objective of anti-poverty policy and is addressed especially in European social policy statements. At the same time, there is a large and inconclusive literature on the intergenerational transmission of inequality (Champernowne and Cowell 1998; OECD 2008). The analysis is designed to give the initial conditions of the poverty inequality. Third, an attempt is to approach processes of transition from the initial conditions to the observed old age poverty. Factors influencing this could be due to decisions of the individual, such as education, choice of occupation or family

---

P. Tinios (✉)
Piraeus University, Karaoli & Dimitriou Street 80, Piraeus 18534, Greece
e-mail: ptinios@otenet.gr

A. Lyberaki and T. Georgiadis
Panteion University of Social and Political Science, Department of Regional and Economic Development, 136, Syngrou Avenue, Athens 17671, Greece
e-mail: antiglib@gmail.com; th.georgiadis@gmail.com

arrangement, patterns of savings; they could be due to unforeseen events or shocks: an illness, family breakup, unemployment, migration; they can finally be due to public interventions in the form of income transfers, both while working and in retirement. Particular attention needs to be paid to key features of the type of social policy of relevance to our cohorts and in operation at the mid-point of our sample's lives: the size of the Welfare State, the emphasis it placed on family, labour and social inclusion and the extent of means testing in operation. Expenditure on pensions and most health care expenditure, directed as they were at the previous cohorts, should be excluded from the picture.

Has globalization led to more poverty? Has the intervention of the Welfare State prevented the emergence of poverty? Have the funds disbursed in the form of social programs bought greater equality, and where? Do our detailed data support Sapir's (2005) observation that in some countries social expenditure does worse both in terms of economic efficiency and equity? Our investigation is motivated by these large questions; it is our hope that some light may be shed on them.

2.2 What Is to Be Explained? Poverty 2004–2007

In European discussions the concept of poverty, albeit with many reminders of its multidimensional nature, has taken centre stage in the attempt to shift emphasis away from discussing efforts at social policy (e.g. by comparing expenditures) and towards outcomes and effectiveness. In political discourse the risk of poverty stands in as the politically most sensitive indicator, a litmus test of efficacy of the "European Social Model", as well as a test case of the negative effects of globalization. The use of the concept may also be sanctioned by the observation that many social scientists stress the existence of cycles of deprivation, which reinforce and make permanent the effects of deprivation. The starting point of the analysis thus is the state of poverty as portrayed in the two waves of SHARE.

The issues raised are discussed in two papers (Lyberaki and Tinios 2005, 2008), so need not be repeated at length: Comparisons are based on net equivalent household income, while the poverty line is corrected for age effects using information about the relative poverty rates for the 50+ population as they are computed using the Survey of Income and Living Conditions (EU-SILC). (See Christelis et al. 2009 on how these variables are defined and how they relate to other features, such as wealth and indebtedness). However, as poverty analysis focuses on the bottom part of the income distribution – on what are by definition extreme observations – it is most likely to be affected by data cleaning, imputations and other technical interventions. The numbers presented in Fig. 2.1, using the imputations from wave 1 and 2 of the SHARE, though generally higher are not unlike what is familiar from other sources such as EU-SILC (Eurostat 2009).

Using information from both previous waves one can construct a new definition of poverty, depending on whether a particular individual in the longitudinal sample is classified under the poverty line in both years, in 1 year only or in none. This measure

2 Explaining Persistent Poverty in SHARE: Does the Past Play a Role?

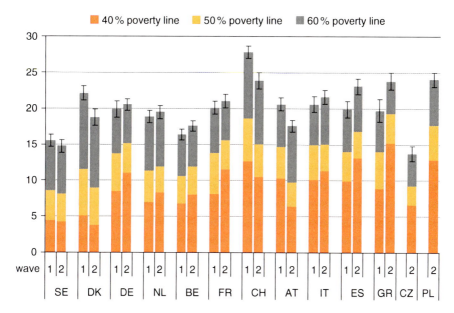

Fig. 2.1 Poverty with alternative lines (age-corrected) Wave 1 and Wave 2

of "persistent poverty" may be interpreted as capturing permanence in poverty (OECD 2008, Chap. 6). Alternatively, given that SHARE's wave 1 collected gross income (corrected for taxes and social insurance contributions in the poverty analysis) and wave 2 focused on net income, the analysis may be given an errors-in-variables interpretation as capturing more closely "true" poverty or the "hard core" of poverty (the population of the poor is mostly below the income tax threshold, so the gross/net distinction would apply to social insurance only). However it may be, Fig. 2.2 shows that there is considerable movement around the poverty threshold for the longitudinal sample.

Persistent poverty is experienced by between 6% (SE) and 14% (GR) of the sample, while a considerably larger proportion (between 21% in SE and 37% in ES) runs the risk of being classed as poor at least once in the two waves. In general, the South experiences higher rates of both poverty and persistent poverty, though the differences are far less pronounced than is familiar from other surveys (e.g. the EU's SILC survey – see Eurostat 2009, Statistics in Focus). In most countries there are more people moving into than moving out of poverty. It is noteworthy that the increase is largest in Spain, where the collapse of housing prices affected imputed housing income for poor owner-occupiers.

What is the subjective significance of poverty, measured as it is conventionally as a property of the shape of the bottom part of income distribution? A note of caution is sounded by the unexpectedly low percentage of persistent poor who state that they can only make ends meet "with great difficulty" (Fig. 2.3). Thus only 6.6% of the persistent poor in SE could not make ends meet in both years, while fully 83% did not mention "great difficulty" at all. Conversely, the far higher percentages

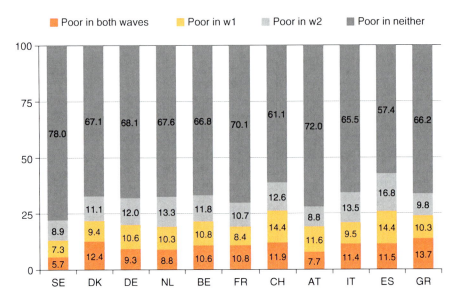

Fig. 2.2 Poverty and persistent poverty in the longitudinal sample (60% median poverty line)

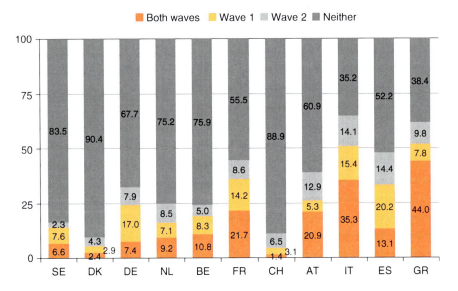

Fig. 2.3 Per cent of persistent poor who make ends meet "with great difficulty"

in the South confirm that poverty is, subjectively, more pervasive and less easily bearable phenomenon than in the North. In GR and IT in each year 50% of that year's poor state they make ends meet with difficulty. Such an observation could be explained by greater stoicism in the North, while it could also be due to benefits in kind making the consequences of a risk of income poverty less important and with fewer consequences for the individuals affected.

The "making ends meet" question as a kind of subjective evaluation of poverty is investigated in detail by Litwin and Sapir (2009) using SHARE wave 1 data. Interestingly, they find that income systematically under-predicts subjective poverty among the very aged, and may even become insignificant for the oldest old. This serves as a reminder that the lived experience of poverty in old age may mean very different things in different countries.

2.3 Initial Conditions: Childhood Deprivation in the SHARE Sample

How far is the poverty we now observe among the older population simply the reflection of inherited deprivation? How far is poverty today conditioned by initial conditions during the childhood of our sample?

SHARELIFE has a number of questions on which a picture of deprivation at age 10 can be built up. Eleven events that serve as indicators of childhood deprivation were selected: (1) Fixed bath in accommodation at the age of 10, (2) Cold running water supply, (3) Hot running water supply, (4) Inside toilet, (5) Central heating, (6) None or very few (0–10 books) in accommodation, (7) Over-crowded accommodation (three or persons per room), (8) Experienced financial hardship during childhood, (9) Experienced hunger during childhood, (10) Poor health status during childhood and (11) A class of origin indicator (breadwinner of the household working as a farmer or in elementary occupation).

Each of these indicators measures absolute deprivation – in the sense of applying a uniform deprivation criterion across all countries and cohorts. What our analysis requires, however, is an idea of relative deprivation (or its opposite, i.e. relative well-being); an impression thus needs to be built up of how each individual stood as compared to what was the norm around him/her. We have thus used Tsakloglou and Papadopoulos (2001) each person's categorical "welfare indicator", which is defined as the sum of the deprivation indicators built from the 11 variables above, weighted by the respective country's proportion being not deprived in the specific indicator (details are available by the authors on request). For each country, the estimated childhood relative well-being index takes for each person values between 0 (complete deprivation) and 1 (no deprivation). In other words, a higher value implies less relative childhood deprivation (more well-being) for a citizen of a country. In essence, this approach assumes that being deprived during childhood of one event, matters more if that event was more widespread, and hence is more likely to be thought to be part of the "social norm". (To ensure adequate degrees of freedom, the analysis was conducted by country, and not by cohort, as logic would imply). Assigning a "deprivation value" to each event depending on its distance from the social norm, has also been employed in other poverty studies (Delhausse et al. 1993).

The results appear in Fig. 2.4. The current persistent poor enjoyed considerably lower relative well-being than those who are not persistently poor. The difference is

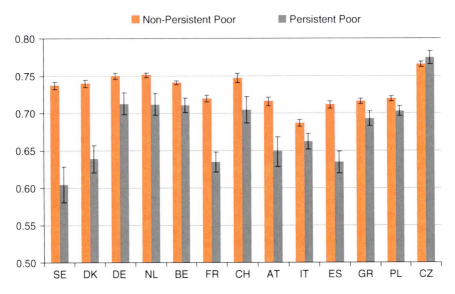

Fig. 2.4 Childhood relative well-being index and current poverty status

largest in the Nordic countries and smaller for the South and is always significant except for the Czech Republic.

How closely does childhood deprivation correlate with poverty in 2004–2007? Figures 2.4 and 2.5 examine the incidence of poverty among the childhood deprived and vice versa. For the purposes of these analyses (and in the absence of wave 1 data), those "poor" in wave 2 in PL and CZ are deemed as "persistent poor".

Figure 2.5 examines the opposite question. How did those who started off with a disadvantage (interpreted as being in the worse-off 20% in their country) fare in their lives? Being deprived at childhood always translates into greater probability of being poor and/or persistent poor –with the notable exception once more of the Czech Republic.

The impression one gleans from this brief tour of childhood deprivation and contemporary poverty is of a strong link between childhood and later life. The cumulative lack of luck to end at the bottom of the income distribution close to the end of one's life is not automatic, but is subject to variation. These variations may be due to own decisions or to chance accidents. All these effects may be mediated by societal structures, primarily of those of social protection. It is to these factors we now turn.

2.4 Transitions to Poverty: A Poverty Probit

Contemporary poverty after age 50 is the cumulative result of decisions and events during the lifetime of the person. Seen in this light it is hardly surprising that no single link or relationship is immediately evident in a preliminary naked-eye analysis.

2 Explaining Persistent Poverty in SHARE: Does the Past Play a Role?

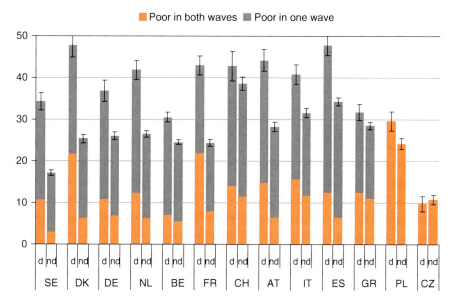

Fig. 2.5 Composition of persistent poverty status, by childhood deprivation status. *Note*: "d" means childhood deprived; "nd" means not deprived in childhood

The effects are likely to be complex and will be mediated through other decisions and through social structures which themselves changed over time. This means that (a) a multidimensional analysis is appropriate in picking up marginal effects and (b) if the focus is on social policy a careful model of Welfare State parameters is essential.

This section proceeds to a probit analysis explaining persistent poverty. A probit may be seen as a partial initial response to the first requirement and possibly could provide a start on the second.

The dependent variable is contemporary persistent poverty (1 if poor, 0 if not). The explanatory variables are divided into 5 categories:

1. Demographics: We distinguish three age cohorts 50–65, 65–80 and 80+. Gender, marital status (widow) and foreign-born.
2. Childhood relative well being index in continuous form (0 = maximum deprivation).
3. Life time experiences: number of children, Inheritance (>5,000 euro), temporary leave of absence for one year or more due to ill-health or disability.
4. Effects of socio-economic decisions: Never in paid work, Type of paid work (elementary-blue collar, professional-clerk), Education, Retirement history (Employed, retired in past 5 years, in the past 6–20 years, more than 20 ago).
5. Characteristics of the Welfare State in the middle-life (decade of their 40s) of the respondent. Three linked indicators are included, drawing from the ESSPROS dataset: (a) overall size as measured by social protection as % of GDP, which includes matters such as education only if they are targeted on

income (e.g. scholarships to poor pupils), (b) Per cent of total social protection expenditure accounted by family, unemployment, housing and social exclusion function (i.e. not pensions or health) and (c) Per cent of total social protection expenditure accounted for by means-tested benefits. Given that all these variables were subject to different interpretations in transition countries, a transition dummy is also included.

Most of the variables have the expected signs and can provide a plausible account of how initial deprivation may translate to current poverty (Table 2.1):

Table 2.1 Determinants of Persistent Poverty Status: Probit Results

Dependent variable = Persistent Poor 2004/2007	Marginal effect	Std error
Demographics		
Age: 50–64 years	0.0645**	0.0100
Age: 65–80 years	F	
Age: over 80 years	−0.0143	0.0114
Female	−0.0115	0.0075
Widowed	0.0417**	0.0115
Foreign-born	0.0613**	0.0186
Childhood deprivation index		
Childhood non-deprivation index: for each country ranges from 0 (complete deprivation) to 1 no deprivation	−0.0885**	0.0234
Life-time experiences		
Number of children	0.0112**	0.0022
Received inheritance (>5,000 euros)	−0.0031	0.0111
Temporary leave of absence from a job for 1 year or more because of ill health/disability	0.0279*	0.0100
Socio-economic characteristics		
Never in paid work	F	
Paid work: Elementary/Blue-collar	−0.0097	0.0112
Paid work: Professional / Clerk	−0.0372**	0.0117
Primary education or lower	F	
Secondary education	−0.0611**	0.0079
Higher education	−0.0731**	0.0084
Other employment status	F	
In employment	−0.0736**	0.0074
Retired in past 5 years	−0.0463**	0.0113
Retired in past 6–20 years	−0.0578**	0.0085
Retired in more than 20 years ago	−0.0404**	0.0091
Social protection expenditure as (%) of GDP in middle-life of the respondent	0.0024*	0.0012
Other social benefits expenditure (except for pensions and health-related benefits) as (%) of social protection expenditure in middle-life of the respondent	−0.0015**	0.0006
Means-tested benefits as (%) of social protection expenditure in middle-life of the respondent	−0.0029**	0.0010
Transition countries (CZ; PL; GDR)	0.1239**	0.0169
Pseudo R-square	0.116	
Number of observations	17,478	

Source: SHARE Wave 1, Wave 2 release 2.3.0; Wave 3, release 0
**, *: Significant at 1%, 5% respectively

The age dummies must be seen in conjunction with the years from retirement variable; being still in employment counteracts a positive effect of being in the younger cohort. Being a widow and being foreign-born exert important influences, as is having had many children or having to have quit a job due to injury. White collar occupations and education predictably play an important role in preventing current poverty.

The childhood well-being index in all cases plays an important and well-defined role, in the sense that its marginal effect appears to depend very little on specification. An interesting observation is that its value alters very little once country dummies and other country specific information is present; this may be taken as evidence that the way original deprivation impinges on current poverty is fairly uniform across countries. If this is so, then it would follow that the type of individual who is most likely to develop from childhood relative to late life poverty is relatively invariant between countries, an important observation in itself.

Turning to the Welfare State characteristics, the emerging picture is very interesting. The proportion of social expenditure spent on matters that affect people of working age (maternity, housing, unemployment) always reduces the probability of being persistent poor in later life, as does the extent of means testing. Well-targeted social expenditure in the mid-life of our respondents continues to pay dividends in later life. The weakly significant positive effect of total expenditure may also have an appealing interpretation: High values for this variable signal large expenditure on old age protection and health. Our respondents at this stage in their lives were mostly working and healthy; for them high pension expenditure would be associated with greater taxes rather than higher benefits. We may interpret this effect as a kind of crowding-out of one kind of social welfare by another. The positive sign of this variable is dependent on the presence of a transition country dummy. Excluding transition countries from the estimation reinforces the welfare state effects noted.

Once country group dummies (roughly corresponding to Esping-Andersen's 1990 groups) are added, the influence of family and other social policies increases, means testing becomes insignificant and total social protection expenditure virtually disappears. Once dummies are included for all countries all Welfare State variables become insignificant, though they retain the original signs.

Thus, welfare policies matter, as does the extent of targeting adopted. However, with minor exceptions, the influence of social policy is benign in a generalized way, affecting the probability of being persistent poor across the board. We have uncovered – at this level of generality at least – little evidence of social policy interacting with specific features of individual lives such as employment or family choices. In particular, the influence of initial conditions is in all cases very strong and it is little altered by adding information on social policy or allowing for specific national effects.

How strong is the social policy effect? To illustrate the answer our probit model gives, at first a baseline high-poverty group was selected and its expected probability of being current persistent poor was computed. Widows, with elementary education, and facing maximum deprivation in childhood, if social policy is at

Table 2.2 The predicted effect of social policy in the probit model, relative to a deprived benchmark

Predicted persistent poverty risk across social policy models	Predicted persistent poverty risk	Relative change (%) vis-à-vis the benchmark
Benchmark: Deprived individual + At the means of social policy parameters (SHARE countries)	28.90	–
Germany's means tested social policy; Denmark's non-pensions and health-related social policy	23.47	−18.8%
Average means tests Denmark's family social policy	24.66	−14.7%
Germany's means tests; Average family social policy	27.60	−4.5%
Greece's means tests; Italy's family social policy	31.91	10.4%
Greece's means tests; Italy's family social policy; Italy's total social expenditure	32.24	11.6%

Notes: Baseline population group (Widowed, with elementary education, non-professional occupation, relative deprived in childhood)

mean levels for the whole sample face a probability of 28.9%. If, however, they lived in a hypothetical country with Germany's targeting and Denmark's family policy, the same group would face a probability of 23.5%, a reduction of 5.4 points or of 18.8%. Conversely, if they were unlucky enough to face Greece's means testing, Italy's family expenditure, and without corresponding reduction in old-age spending (Italy's total expenditure), the predicted probability would rise to 32.2%, a 3.3 points rise or 11.6 relative to the benchmark. The range from the worst social policy case to the best one is from 23.5% to 32.24, fully 8.7% points, close to the overall mean. Table 2.2 shows these and other combinations of social policy stances.

The effect of original deprivation was gauged relative to the same benchmark by simulating the effect of moving the deprivation index from its minimum (in the baseline) to the median value of the deprivation index for all countries. The baseline probability drops drastically from 28.9% to 21.4%, a reduction of 7.5 points, a little less than the range due to social policy.

2.5 Conclusion: A "European Social Model"?

The results of this paper appear to affirm that "social policy pays". Not only contemporaneously as a palliative of immediate problems, but also as a long-term investment. Our results indicate that, once original deprivation is taken into account, targeted public expenditure on family, housing, labour and social exclusion continues to pay dividends *decades* later – when the beneficiaries have reached retirement age and onwards.

This paper began by asking a number of questions about the effects and efficacy of social policy. Those who think of social expenditure as "a factor of production", may find some corroboration for their position; the reach and effectiveness of this factor, though, was by no means uniform. The search for the European social model can go on.

References

Champernowne, D. G. & Cowell, F. A. (1998). *Economic Inequality and income distribution.* Cambridge.

Christelis, D., Japelli, T., Paccagnella, O., & Weber, G. (2009). Income, wealth and financial fragility in Europe. *Journal of European Social Policy, 19*, 359–376.

Delhausse, B., Luttgens, A., & Perelman, S. (1993). Comparing measures of poverty and relative deprivation: An example from Belgium. *Journal of Population Economics, 6*, 83–102.

Esping-Andersen, G. (1990). *The three worlds of welfare capitalism.* Cambridge: Polity press.

Eurostat (2009) *Statistics in Focus, number 46, "79 million EU citizens were at risk-of-poverty in 2007 of whom 32 million were also materially deprived".* Luxembourg.

Litwin, H., & Sapir, E. (2009). Perceived income adequacy among older adults in 12 countries: Findings from the Survey of Health, Ageing, and Retirement in Europe. *The Gerontologist, 49*, 397–406.

Lyberaki, A. & Tinios Pl. (2005). Poverty and social exclusion: A new approach to an old issue. In A. Börsch-Supan, A. Brugiavini, H. Jürges, J. Mackenbach, J. Siegrist & G. Weber (Eds.), *Health ageing and retirement in Europe: first results from the Survey of Health, Ageing and Retirement in Europe, Mannheim* (pp. 302–309).

Lyberaki, A & Tinios Pl. (2008) Poverty and persistent poverty: Adding dynamics to familiar findings. In A. Börsch-Supan, A. Brugiavini, H. Jürges, A. Kaptein, J. Mackenbach, J. Siegrist & G. Weber, *Health ageing and retirement in Europe (2004–2007): Starting the longitudinal dimension, Mannheim* (pp. 276–283).

OECD (2008). *Growing unequal? Income distribution and poverty in OECD countries.* Paris.

Sapir, A. (2005). *Globalisation and the reform of European social models, Bruegel Policy Contribution.*

Tsakloglou, P. & Papadopoulos F. (2001) Indicators of social exclusion in EUROMOD, EUROMOD, Working Paper No. EM8/01.

Chapter 3
Childhood, Schooling and Income Inequality

Danilo Cavapozzi, Christelle Garrouste, and Omar Paccagnella

3.1 Childhood Effects Over the Life-Cycle

Parental socio-economic background plays an important role in determining employment outcomes during the individual's whole life-cycle. Indeed, the environment in which individuals grow up plays a crucial role in determining their later socio-economic condition, regardless of their own abilities. On the one hand, this link might be due to the transmission of social norms (e.g. work ethics) or risk attitudes from parents to children. On the other hand, public policies may strengthen or weaken cross-generational persistence in the socio-economic status. For instance, whereas elitist education systems support the assumption of innate abilities and therefore focus their efforts and resources on the most promising pupils, redistributive policies supporting access to education at all levels (i.e. egalitarian education systems) are a typical example of interventions aimed at loosening the dependence between human capital accumulation and parental background. After the Second World War and up to the early 1980s, European education systems were demonstrating high within-country variance in learning achievement (e.g., Foshay et al. 1962). Progressively, most countries did, however, move from elitist systems to more egalitarian systems (EGREES 2003).

D. Cavapozzi (✉)
Dipartimento di Scienze Economiche "Marco Fanno", Università di Padova, Via del Santo 33, 35123 Padova, Italy
e-mail: danilo.cavapozzi@unipd.it

C. Garrouste
European Commission – Joint Research Centre (EC – JRC), Institute for the Protection and Security of the Citizen (IPSC), Unit G.09 Econometrics and Applied Statistics (EAS), TP361 – Via Enrico Fermi 2749, 21027 Ispra (VA), Italy
e-mail: christelle.garrouste@jrc.ec.europa.eu

O. Paccagnella
Dipartimento di Scienze Statistiche, Università di Padova, Via Cesare Battisti, 241/243, 35121 Padova, Italy
e-mail: omar.paccagnella@unipd.it

Our analysis exploits the richness of the SHARELIFE data on household economic resources and cultural background of respondents at the age of 10 to address two main research questions. First, we want to show how disparities in the financial and cultural background during childhood are correlated with disparities in education and income at the first job. We focus on educational attainment and incomes at the beginning of the working career because these outcomes are more likely to take place at a stage of the life-cycle relatively close to childhood and youth and to reflect the influence of parental household status. Second, we will study the correlation between income inequality at first job and education controlling for childhood background in order to evaluate to what extent policies promoting education access and longer educational attainment can be of use to reduce income dispersions at the beginning of the working career, once childhood disparities are taken into account. Finding that further full time education is associated with lower income inequality, even for those grown up in disadvantaged households, might support the hypothesis that financing education access at all levels can be an effective strategy to lower income dispersion and finalize the policy goal set by the European Union to decrease income disparities by the next decade (European Commission 2010).

The paper is organized in four sections. While Sect. 3.2 presents cross-country heterogeneity in childhood background, the empirical correlation between childhood and respondents' socio-economic outcomes is investigated in Sects. 3.3 and 3.4 respectively. Finally, Sect. 3.5 summarizes the main findings of our analysis and discusses how public policies contribute to explain cross-country differences.

3.2 Childhood Socio-Economic Indicators

We take advantage of the SHARELIFE questionnaire to calculate the number of rooms per capita in the (private) accommodation where respondents lived at the age of 10. According to the questionnaire, the number of rooms considered in this analysis excludes kitchen, bathrooms and hallways, but includes bedrooms. We opt to consider rooms per capita instead of the overall number of rooms in the accommodation because this latter index is clearly affected by between and within country variability in household size, which limits its comparability. The rooms-per-capita indicator circumvents this problem and is expected to be a more adequate proxy of the level of household economic resources available during respondents' childhood. Indeed, lower levels of rooms per capita suggest the presence of overcrowding in the accommodation, which is an indicator of financial distress. To this end, it is worth noting that we found a strong and positive correlation (correlation index = 0.82) between our rooms-per-capita index and the OECD average disposable income of households with children aged 0–17 (OECD, 2008). This evidence supports our claim that rooms-per-capita is a sound indicator of parental financial status during childhood years.

Denmark, Belgium and Switzerland are the countries associated with the highest relative provision of rooms when the SHARELIFE respondents were 10 years old. In these countries, the average number of rooms per capita is around 0.90, i.e. each household member had almost one room at her disposal. On the contrary, Eastern Europe and Mediterranean countries present the lowest level of room provision. Noteworthy, while in Poland the number of rooms per capita is 0.39, in the Czech Republic and Mediterranean countries it is, on average, about 0.55.

The second indicator is obtained by the SHARELIFE question that asks respondents to provide an estimate of the number of books available in their accommodation at the age of 10. Book availability is measured in terms of number of shelves and bookcases that can be filled. Magazines, newspapers and school books are not considered. This indicator is expected to be correlated with the cultural background of the household where respondents grew up. This reasoning is confirmed by the strong and positive relationship (correlation index = 0.66) existing between number of books at home and country averages for years of education in the adult population reported by Barro and Lee (2000). Then, the higher the number of books, the higher is the expected average educational level of parents, siblings and other relatives in the household.

The original information collected by the SHARELIFE questionnaire is rearranged to discriminate between respondents who had enough books to fill one bookcase and those having fewer books in their accommodation. On average, two thirds of respondents declare fewer books than the equivalent of one bookcase, but this overall result hides relevant cross-country variability. We notice that Mediterranean countries and Poland are by far the countries where books were less widespread in parental respondents' houses (more than 80% of respondents in these countries declare scarcity of books). In particular, this proportion is almost 90% for Italians and Greeks. On the contrary, only 40% of Swedes, Danes and Czechs spent their childhood in a place with few books. Notably, the pattern found for Czech Republic is at odds with that characterizing Poland. In the former country, individuals growing up in an environment with few books are less than one half than those in the latter. Among the remaining countries, respondents living in Belgium, France or Austria are those who experience the highest book-shortage during childhood.

Overall, our indicators show relevant cross-country heterogeneity in room provision and number of books in the accommodation of respondents at the age of 10. We chose to base our study on such indicators since both of them explicitly refer to the same stage of respondents' childhood (age of 10) and the information conveyed by them can easily be compared across respondents and countries. In fact, the concepts underlying rooms per capita and number of books are objective and cannot be altered by individual reporting styles due to, for instance, cross-country heterogeneity in tradition and social norms.

The next step of our analysis is to look at the correlations between these childhood indicators and the number of years spent in full-time education by respondents. Assuming that rooms per capita and number of books are good predictors of financial status and cultural background of parental household, such analysis will help understanding whether inequalities in the resources available during childhood end up in significant differences in individuals' educational attainments.

3.3 Childhood Resources and Educational Attainments

As discussed above, the number of rooms per capita is correlated with household financial status and might be of help to discriminate between better off and worse off households. To address this issue, we split respondents in two groups: in the former we group all individuals reporting a number of rooms per capita lower than the 25th percentile of their country, the latter includes those with a higher level of relative room provision. We used country-specific percentiles to allow for cross-country heterogeneity in overall macro-economic conditions and accommodation arrangements. Given the positive correlation between rooms per capita and average disposable income, individuals who are at the bottom of their country ranking of rooms per capita are expected to be those living in worse off households.

Figure 3.1 shows the average number of years spent in full time education by country and by relative level of room provision. In our analysis time spent in full time education is calculated taking into account the fact that the age when starting schooling varies over time and across countries. Moreover, it should be noted that the evidence in this section shows the relationship between childhood background and education controlling for gender and birth-cohort heterogeneity. On average, individuals with lower levels of rooms per capita spent about 2 years less in full time

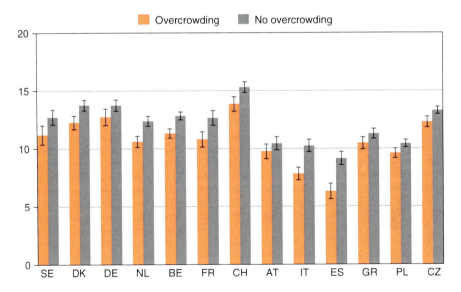

Fig. 3.1 Years of full time education, by country and rooms per capita at the age of 10

3 Childhood, Schooling and Income Inequality

education than those who lived in an accommodation with more rooms per capita. This difference, although small on average, is statistically significant in most countries and suggests that individuals with better off parental households have spent a higher number of years in full time education. The widest differentials are found for Italy and Spain, where living in accommodations with a lower number of rooms per capita is associated with a reduction in the full time education period by more than 2 years.

Figure 3.2 shows the variation across countries in the average number of years spent in full time education by availability of books in respondents' accommodation. Since the number of books in the accommodation may increase with the number of household members, in addition to gender and year of birth fixed effects, we also control for household size. On average, respondents who have more books at their disposal remain in full time education 3 years longer. This variation is statistically significant.

On the one hand, Denmark, Switzerland and the Czech Republic are the countries where the variation is the lowest and amount to less than 2 years. On the other hand, education differentials are the widest for France, Italy and Spain. In particular, for Italy it amounts to almost 6 years on average. Hence, cultural background affects significantly the amount of full-time education received in all countries. Better educated households are associated with a higher educational attainment. As before, the strength of this link exhibits clear cross-country variations. Again, the relationship between education outcomes and cultural background of household members appears more pervasive for Italy and Spain. The results from Figs. 3.1 and 3.2 were confirmed when re-running the regression considering the variability of the control factors both by country and by group.

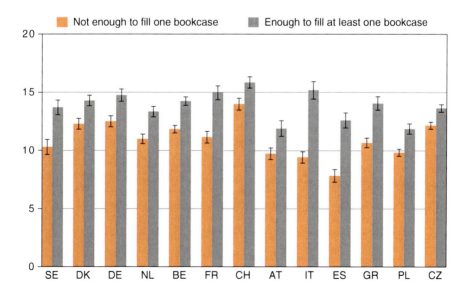

Fig. 3.2 Years of full time education, by country and number of books in the accommodation at the age of 10

This evidence, overall, reconciles with what we found earlier when investigating the relationship between education and the financial status of parental household. Institutions of Italy and Spain seem to be those less able to encourage intergenerational mobility and loosen the influence of parental socio-economic background on children outcomes. These within- and between-country differences confirm the findings from the first 12-country study of 1959–1961 by the UNESCO Institute for Education (Foshay et al. 1962). Comparing internationally standardized scores of 13-year-olds from mathematics, non-verbal, reading, geography and science tests, that pioneer study revealed a difference of roughly three-fourths of a standard deviation in average score between the father's lowest educational category and the highest, and a range of about two-thirds between extreme father's occupational groupings.

3.4 Childhood Resources and Income Inequality

In this section we analyse how respondents' incomes are related to the socio-economic status of their parents and how education attainment affects the reproduction of that socio-economic status. Investigating these relationships is particularly relevant in the European context as the reduction of income disparities has become one of the five headline targets for the EU27 to be reached by 2020, alongside with higher educational attainment (European Commission, 2010).

As demonstrated by the literature, the relationship between the distribution of incomes and the distribution of educational attainment is indeed twofold: first, the presence of income inequality during childhood may prevent access to education when education is too costly and may prevent equal achievement when it is combined with socio-cultural inequality; second, improved access to education is expected to raise the earning opportunities of the lower strata and, ceteris paribus, reduce earnings inequality (Mayer 2010). Hence, policy measures easing access to education constitute a key tool against a self-perpetuating poverty trap.

In this analysis we focus on the cohort of respondents born in 1940 or later who had at least one paid job (as employee or self-employed) and report their initial income from work after taxes. It is worth noting that considering only respondents born in 1940 or later leads to exclude from the analysis about one third of the original sample but has the advantage of partially controlling for cohort effects because we consider only the individuals who grew up in the period following the World War II. Also, since we are focusing on individuals having at least one employment spell, we dropped all the respondents in this so-selected sample who have never worked in their life (about 6.5%). For each country we consider only the amounts expressed in euro or pre-euro local currency and exclude individuals who do not report usable information on the amount and the currency of their first income from work as well as for the year in which their first employment spell started. Within-country revaluations and devaluations of local currencies have been taken into account. We use consumer price indices (base 2006) to control for time-varying inflation rates and calculate amounts in real terms. Since product price formation was centrally controlled in all countries of the Soviet bloc until the 1990s, we exclude amounts expressed in East Germany marks and individuals living in Poland and Czech Republic. Our final sample consists of about 8,200 observations.

We measure income inequality by means of the interquartile range (IQR) to the median ratio (Christelis et al. 2009). Specifically, the IQR is defined as the difference between the 75th income percentile and the 25th income percentile. We opt to consider this relative measure of income inequality rather than the standard coefficient of variation (standard deviation to mean ratio) because median and interquartile ranges are more robust to outliers in the distribution tails than means and standard deviations. Outliers might produce spurious heterogeneity in income amounts due to memory effects that affect respondents' answers when they are asked to recall events occurred in the past. We point out that the IQR to the median indicator can by construction be directly compared across countries, since it overcomes the problem of the currency in which amounts are expressed.

We look at the relationship between income inequality at first job and number of books in childhood accommodations. First, we notice that median first income for respondents with scarcity of books tends to be lower than median first income for respondents with a higher number of books. Figure 3.3 shows how the IQR to median ratio indicator for the first job income varies with the availability of books in the parental accommodation. As expected, we find that countries with lower proportions of individuals grown up in households with few books are those with lower income inequality. Moreover, the gender disaggregation reveals a weaker correlation between the number of books and income inequality for women than for

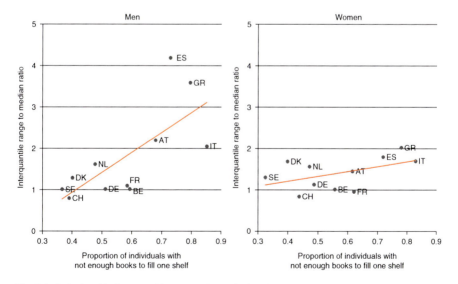

Fig. 3.3 Relationship between IQR to median ratio for first job incomes and number of books in the accommodation at the age of 10 by gender

men and less variability of the IQR to the median. Regardless of the gender, Scandinavian countries, Switzerland, Germany and the Netherlands report both lower proportions of individuals with few books in their accommodation at the age of 10 and lower income dispersion around the median. Vice versa, Mediterranean countries report more individuals with few books at home during childhood and higher income dispersion.

We now consider the relationship between first job incomes and number of rooms per capita in parental houses at the age of 10. As before, the median first income for respondents who lived in an overcrowded house is overall lower than the median first income for respondents who did not live in an overcrowded house. Figure 3.4 shows the variations across countries of the IQR to median ratio for the first job income by gender and rooms per capita at the age of 10. Results are comparable to those of Fig. 3.3: countries where individuals grew up in accommodations with lower number of rooms per capita are those with the highest income inequality. This suggests that countries with lower disparities in the economic resources available during childhood are also those where employment and self-employment income at the first job is more concentrated around the median. Remarkably, we noticed that the North-South gradient reported in Fig. 3.3 is entirely confirmed in Fig. 3.4. Further, we still find a significant difference across genders with a much stronger correlation between IQR and childhood housing conditions for men and less dispersion for women.

The findings presented so far in this section suggest that the childhood background is significantly correlated with income inequality at first job. Overall, countries where individuals are more likely to grow up in better-off and better-educated households always show lower income inequality. Income inequalities

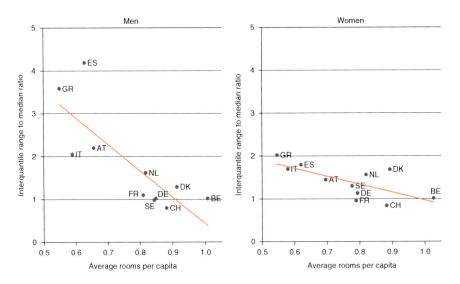

Fig. 3.4 Relationship between IQR to median ratio for first job incomes and average rooms per capita by gender

are, however, even more affected by the childhood factors in the case of men than in the case of women. Cross-gender differences in this relationship as well as income inequality levels may result from cross-gender heterogeneity in labour market participation and the relative homogeneity in the types of jobs women attended compared to men. To this end, it is worth noting that ILO statistics confirm both these arguments. They show that women at work between 1945 and 1970 constituted, on average, 26% of the total labour force and that they were mostly concentrated in services and industry sectors.

Given the relationship between childhood background and years of education (cf. Figs. 3.1 and 3.2), we expect lower income inequality in countries where individuals spent on average more years in full-time education.

Figure 3.5 confirms this pattern. On average, the higher the number of years of education, the lower is the IQR to median ratio for first incomes, with again a lower dispersion for women than for men. Such country level information is used to calculate the percentage variations in the income dispersion at first job with respect to a marginal change in the years of education. An increase in the average time spent at school by 1 year is correlated with a reduction in the IQR to median ratio by 28% for males and 15% for females. Although these results confirm the negative association between education and income inequalities, they do not tell how this relationship is affected by family background. Indeed, it can be argued that countries where individuals spend on average more years in full-time education are also those where children are more likely to grow up in better educated and richer households. In other words, the positive correlation between the socio-economic condition of the parental household and educational attainments might

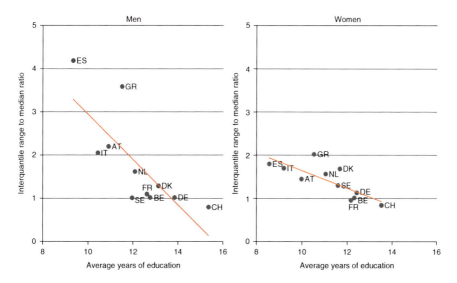

Fig. 3.5 Relationship between IQR to median ratio for first job incomes and average years of education by gender

drive the evidence reported in Fig. 3.5 and lead to misleading interpretations on the association between education and income inequality.

Therefore, to address this issue we study the relationship between income inequality and education disaggregating the sample by gender and childhood background in terms of number of books and rooms per capita in the accommodation at the age of 10. Overall, our results show that income inequality and education are negatively related even controlling for parental household characteristics. Thus the evidence in Fig. 3.5 seems not to be driven by heterogeneity in the sample composition due to childhood background disparities.

In particular, Figs. 3.6 and 3.7 respectively illustrate our findings for individuals from low economic background (i.e. crowded housing condition) and from low socio-cultural background (i.e. a number of books not sufficient to fill one bookcase). Regardless of the childhood indicator considered, there is a clear negative relationship between average years of education and income inequality for both men and women. Higher education is associated with lower income dispersion even if individuals grew up in socio-economically disadvantaged households. If we focus on males experiencing overcrowding or book shortage at the age of 10, a marginal increase in the average years of education is associated with a decrease in the IQR to median ratio by 11% and 23% respectively. For females, these reductions are equal to 14% in both cases. Although these estimates do not describe causal effects, they stress the strong relationship existing between income variability and educational attainments in our sample.

Our descriptive evidence clearly suggests that public policies fostering education access and extending the length of full time educational attainment can be an effective strategy to decrease income inequalities in Europe. To provide a further

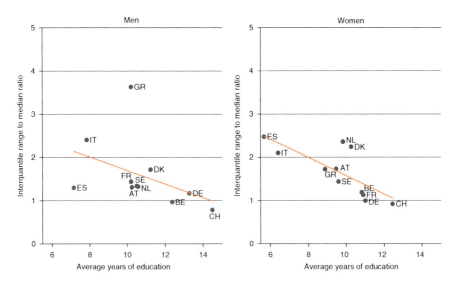

Fig. 3.6 Individuals experiencing overcrowding in the accommodation at the age of 10: relationship between IQR to median ratio for first job incomes and years of education by gender

3 Childhood, Schooling and Income Inequality   41

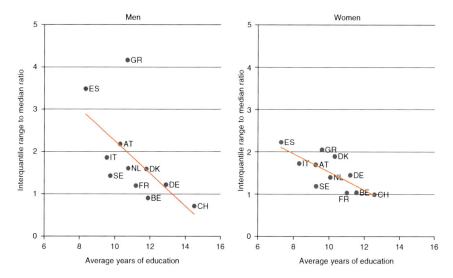

Fig. 3.7 Individuals experiencing book shortage in the accommodation at the age of 10: relationship between IQR to median ratio for first job incomes and years of education by gender

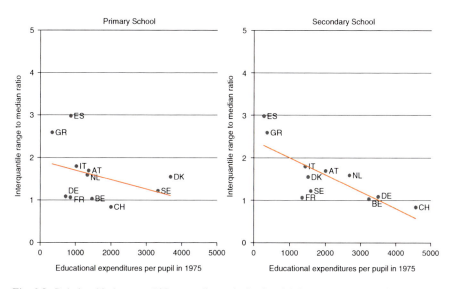

Fig. 3.8 Relationship between IQR to median ratio for first job incomes and educational expenditures per pupil at primary and secondary school

underpinning to this interpretation, Figure 3.8 crosses our measure of income inequality with contextual data on public educational expenditures per pupil at primary and secondary schools from 1975 (Barro and Lee 1997). All amounts are expressed in PPP-adjusted 1985 international dollars and make it possible to draw meaningful cross-country comparisons. Figure 3.8 shows that higher amounts of resources devoted to primary and secondary education are associated with lower income inequality at first job. Notably, in both cases Italy, Spain and Greece combine high levels of income inequality with low educational expenditures per pupil. Hence, this later result comforts our assumption that public intervention in education access may play an important role in explaining income inequalities at the beginning of the working career.

3.5 Discussion and Concluding Remarks

This contribution provides descriptive evidence that financial and educational background of parental household plays an important role in determining individual socio-economic outcomes. Educational attainments (years of schooling) and income inequality are shown to vary with the environment in which individuals grew up. In particular, respondents living in better off and better educated contexts, on average, remain in full time education longer and exhibit lower income inequality. This pattern is present in all countries and it is found to be particularly pervasive for Mediterranean countries which associate poorer socio-economic conditions during childhood with higher differentials in years of full-time education and higher income inequality.

Education policies may play a role in explaining these observed differences across countries. The introduction of support systems fostering access to education of students from disadvantaged households can weaken their financial dependence from parents and loosen the persistence in socio-economic conditions across generations (Mayer 2010). Typical examples of these policies are government supported loans or grants to finance higher studies. OECD analyses (OECD 2010) show that in countries where support programs are available students from worse off households have a lower gap in the probability of attaining tertiary education levels with respect to their counterparts living in high income households. Indeed, this evidence is in line with our results showing that education differentials associated with childhood background are larger in Mediterranean countries, where no universal funding systems were available at the time of schooling of our sample, than in Denmark, where some forms of tuitions or grants were already experimented for pupils from more disadvantaged background (Garrouste 2010).

In addition, our analysis shows that the countries where individuals remain in full-time education longer present lower income disparities and that this negative correlation is confirmed even after controlling for childhood background. This amounts to say that, once we compare individuals growing up in similar childhood environments, we still find a negative relationship between education and income inequality at the first job. Conditioning on childhood socio-economic status reveals that this pattern is not driven by heterogeneity in the sample composition assigning individuals coming from more disadvantaged households to lower educational attainments and higher variability in their income. Thus, public policies fostering education access and increasing the number of years spent in full time education may qualify as a possible strategy to reduce income dispersions at entrance to the labour market.

References

Barro, R. J., & Lee, J. -W. (1997). Schooling quality in a cross section of countries. NBER Working Paper No. W6198, September.

Barro, R. J., & Lee, J. -W. (2000). International data on educational attainment: updates and implications. CID Working Paper No. 042, April.

Christelis, D., Jappelli, T., Paccagnella, O., & Weber, G. (2009). Income, wealth and financial fragility in Europe. *Journal of European Social Policy, 19*(4), 359–376.

EGREES. (2003). *Equity of the European educational systems: a set of indicators*. Liège: Department of Theoretical and Experimental Education, University of Liège.

European Commission (2010). Europe 2020 Plan. available at URL:http://ec.europa.eu/news/economy/100303_en.htm)

Foshay, A. W., Thorndike, R. L., Hotyaat, F., Pidgeon, D. A., & Walker, D. A. (1962). *Educational achievements of thirteen-year-olds in twelve countries*. Hamburg: UNESCO Institute for Education, International Studies in Education.

Garrouste, C. (2010). *100 Years of educational reforms in Europe: a contextual database*. Luxembourg: Publications Office of the European Union, EUR 24487 EN.

Mayer, S. E. (2010). The relationship between income inequality and inequality in schooling. *Theory and Research in Education, 8*(1), 5–20.

OECD (2008). *Growing unequal – income distribution and poverty in OECD countries*. Paris: OECD Publications.

OECD (2010). *Economic policy reforms: going for growth*. Paris: OECD Publications.

Chapter 4
Human Capital Accumulation and Investment Behaviour

Danilo Cavapozzi, Alessio Fiume, Christelle Garrouste, and Guglielmo Weber

4.1 Limited Participation in Financial Markets

Limited use of financial markets is associated with financial distress later in life (Angelini et al. 2009). Such limited use may be the result of choice, or, more likely, it may be due to some impediment.

Limited participation in financial markets may be related to transaction costs or high risk aversion, but has also been found to be related to a low level of financial literacy. A number of papers have used explicit and quantitative measures of financial literacy and related them to individual financial decisions (e.g., Guiso and Jappelli 2008; Lusardi 2008). Lusardi and Mitchell (2007) show that more "financially literate" individuals are more "retirement ready". It has also been shown that stock holdings are much less common among the less financially literate (Christelis et al. 2010; McArdle et al. 2009).

Although no direct measure of financial literacy is as yet available in SHARE, we have information on education-related variables that are well known to correlate with financial literacy. In particular, the third wave of SHARE records information not only on educational attainment, but also on self-assessed mathematical ability while at school and on access to books at age ten. Crucially, we also know when an individual first entered risky financial markets, by investing in stocks, mutual funds, individual retirement plans or life insurance policies.

This chapter is divided in five sections including the current introduction. The second section presents the distribution of the types of investment in risky assets

---

D. Cavapozzi (✉), A. Fiume, and G. Weber
Dipartimento di Scienze Economiche "Marco Fanno", Università di Padova, Via del Santo 33, 35123 Padova, Italy
e-mail: danilo.cavapozzi@unipd.it; alessio.fiume@unipd.it; guglielmo.weber@unipd.it

C. Garrouste,
European Commission – Joint Research Centre (EC – JRC), Institute for the Protection and Security of the Citizen (IPSC), Unit G.09 Econometrics and Applied Statistics (EAS), TP361 – Via Enrico Fermi 2749, 21027 Ispra (VA), Italy
e-mail: christelle.garrouste@jrc.ec.europa.eu

and the entry age into these financial markets across Europe; the third section introduces the different human capital components retained for our analysis. The fourth section presents the results of the multivariate duration analysis conducted to investigate the relationship between investment behaviour and human capital accumulation. Finally, the fifth section draws the main conclusions of our analysis.

4.2 When Did Older Europeans First Invest?

It is a well known fact that large fractions of individuals do not own risky financial assets, thus foregoing the extra returns that these assets have so far produced in the long run. The third wave of SHARE asks whether individuals ever owned stocks, mutual funds, individual retirement accounts or life insurance policies. If the answer is yes, respondents are asked to report when they first bought them. The fractions of individuals who answer yes is relatively small overall (24% for stocks, 22% for mutual funds, 26% for individual retirement accounts, 31% for life insurance policies), but varies dramatically across countries. For instance, over 50% of Swedes and Danes report holding stocks at some time over their life course, less than 15% do the same in Austria, Italy, Spain, Greece and Poland. Mutual funds investment follows a similar pattern – with lower participation rates in most countries (Sweden, Germany, Belgium, France and Switzerland are the exceptions). Investment in individual retirement accounts instead reflects much more special features of pension legislation: we observe high values in the Czech Republic (56%), Sweden (53%), France (49%), Belgium (46%) and Switzerland (42%), and low or very low participation rates in all remaining countries. Finally, life insurance policies are wide-spread in Sweden, Germany, and Austria (all above 40%), and rather common also in Poland (31%), where they were quite wide-spread under the communist regime. Life insurance policies are sometimes required by lenders when mortgages are taken up. For this reason, we have excluded them throughout the analysis if their purchase coincides with the purchase of a home (± 1 year around it).

Figure 4.1 plots the fractions of individuals who ever had at least one of the above forms of risky financial investment at some stage in their lives. This fraction is highest in Sweden (91%) and Denmark (80%). It also exceeds 60% in Switzerland, Germany, Belgium, France and the Czech Republic. The lowest values are recorded in Greece (11%), Spain (25%), Italy (28%) and Poland (34%).

Participation rates differ by gender: overall, 60% of men and 47% of women ever invested in at least one risky financial asset. Gender imbalance is lowest in Sweden (93-90), the Czech Republic (63-59) and Poland (37-31), highest in absolute terms in Germany (82-63), the Netherlands (64-42) and Switzerland (84-63). Of the remaining countries, Spain (31-21) and Greece (15-7) have lower differences, partly because both sexes exhibit low participation. These gender differences in investment behaviour are confirmed by other studies which highlight the overall higher risk-aversion of women (e.g., Hira and Loibl 2008).

4 Human Capital Accumulation and Investment Behaviour 47

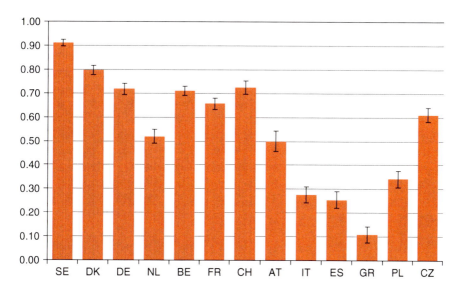

Fig. 4.1 Fraction of individuals who report ever investing in at least one risky financial asset

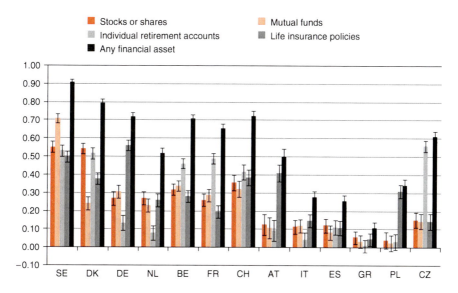

Fig. 4.2 Fraction of individuals who report ever investing in risky assets by asset type

Participation rates are also driven by institutional factors. For instance, the high participation rate of Czech respondents is largely driven by their retirement accounts (which have been heavily tax-subsidized since 1995): only 23% report ever having held stocks or mutual funds. As displayed in Fig. 4.2, the exclusion of retirement accounts has otherwise minor effects on participation. It is worth stressing that in some countries (like the Netherlands or Sweden) individuals have a say in the way their occupational pensions are invested, very much like in an individual retirement account.

To better understand the above distribution of risky financial assets among older Europeans it is useful to investigate at what age the first investment was made. Figure 4.3 displays the age distribution of first-time investment in risky financial assets. The left panel excludes life insurance policies. Not surprisingly, most respondents report starting participation at age 40 or above, with a peak at 50. However, there are non-negligible fractions of first-time investors at younger ages. Given that our respondents were all at least 52 years of age at the time of the interview, those who first invested at ages below 35 must have entered the market before the 1990s. The right panel includes life insurance policies: in this case, investment early in life is much more common. Indeed, some individuals report investing in their teens (possibly for tax avoidance purposes). Note that mortgage-related life insurance policies are not counted here (see above).

Figure 4.4 provides details on the age distribution of financial markets entry by country. For each country, it displays the median, the first and third quartiles, the minimum and maximum. Life insurance policies are excluded in the left panel, included in the right panel. We see in the left panel that the median is lowest in Denmark and France, followed by Belgium, Sweden and Switzerland. France is the country with the largest dispersion (as proxied by the interquartile range), Poland and Greece with the lowest. In interpreting these statistics one has to keep in mind that they refer to the density functions conditional on participation: if marginal investors tend to enter later in life, the higher participation rate in Sweden (over 80%) may explain why the median age of entry is higher than in France, where

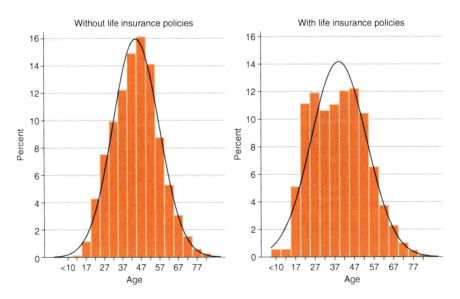

Fig. 4.3 Fraction of individuals who first invested by age. Life insurance policies excluded in the *left panel*, included in *right panel*

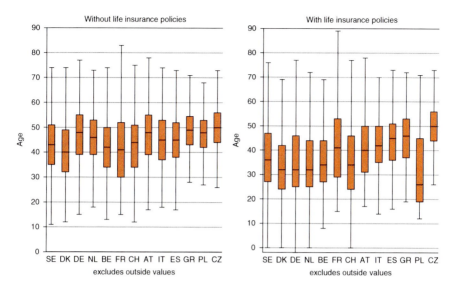

Fig. 4.4 Fraction of individuals who first invested in financial markets, by age and country

fewer participate (less than 60%). The right panel shows that the highest dispersion is attained in Poland, followed by Germany, Switzerland and the Netherlands, highlighting the importance of life insurance policies in promoting early entry into financial markets.

4.3 The Role of Human Capital Accumulation

Human capital is one of the most important determinants of investment decisions. Indeed, higher human capital endowments may lead to a better understanding of financial market opportunities and help individuals in realizing efficient portfolio allocations (Christelis et al. 2010). SHARELIFE data make it possible to analyze human capital along several dimensions (i.e. formal education level, mathematical skills and cultural background) and are of use to investigate whether the cross-country heterogeneity in financial market participation discussed in the previous section can be associated with cross-country heterogeneity in human capital accumulation.

First, we consider the number of years of education, as a proxy of formal education level. Our analysis makes use of context information to take into account that primary school ages vary over time and across countries. As reported in Fig. 4.5, we find much lower median education levels in Spain, Italy, Austria and Greece. Respondents in Denmark, Germany, Belgium, Switzerland and the Czech Republic remain in full time education longer. The greatest variability occurs in Sweden, Greece, Italy and Denmark – by far the lowest in the Czech Republic.

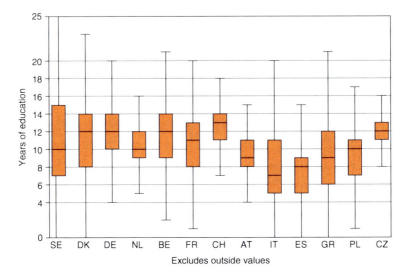

Fig. 4.5 Years of full time education by country

These patterns largely reflect the education reforms on duration of compulsory schooling that took place over the past century (Garrouste 2010), inequalities in the economic resources available during childhood but also individual choice and society attitudes towards education. For instance, median years of education tend to be lower for women in all countries, but in some countries gender differences are particularly strong: we find that in Denmark, Switzerland, Greece and Italy the median number of years of education is lower for women by at least 2 years.

Although there is no doubt that the number of years of full-time education is a key indicator of human capital, the understanding of economic and finance concepts required to manage risky assets is more strictly related to mathematical skills (McArdle et al. 2009). To this end we take advantage of the SHARELIFE question asking respondents to evaluate their mathematical skills at the age of ten as compared to their class-mates. Such self-assessment to some extent reflects individuals' self-confidence. We find that, overall, while more than 30% of individuals rank themselves better in mathematics than their class-mates, less than 15% declare themselves worse. According to the majority of respondents, their understanding of mathematics was about the same as that of their class-mates.

In previous waves of SHARE respondents provided answers to four numeracy questions that help us assess their current mathematical skills in a less subjective way. In our analysis we therefore combine the information on years of education, self-assessed mathematical skills at the age of ten and the number of correct answers to numeracy questions. It is interesting to see how these variables covary across countries. In Fig. 4.6 we report the average numeracy score versus the average self-assessed mathematical abilities at age 10 by education and country. This figure reveals that there is a strong positive relation between mathematical skills at age 10 and educational attainment, but also that current numeracy is

4 Human Capital Accumulation and Investment Behaviour

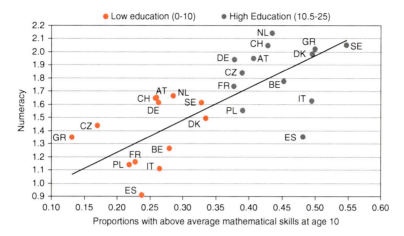

Fig. 4.6 Average current mathematical ability versus mathematical skills at age 10, by education and country

positively correlated both with early life mathematical skills and with formal education. It also shows that on average, in some countries, notably Italy and Spain, but also Poland and France, current numeracy is lower after controlling for education. This may reflect more limited use of mathematical skills in adult life or the type of formal education received. Moreover, it is worth highlighting that the 1960s were marked by an overall international progressive shift towards constructivist mathematics curricula, i.e. a shift in focus from memorization to understanding, launched in the United States. Yet, whereas Switzerland and northern Europe (including Sweden, Denmark and the Netherlands) adopted the new approach promptly, the southern and eastern Europe kept on with the traditional memorization approach until the mid-1980s (Garrouste 2010). Hence, the dispersion of the countries along the correlation trend line (Fig. 4.6) may also be affected by the differences in teaching methods of mathematics at the age of 10.

Finally, we consider the cultural background of parental household by looking at the number of books in the accommodation at the age of 10. This is expected to be an indicator of the average education level of household members and also of respondents' educational attainments. A wide research strain (OECD 2010) shows that there is persistence in the socio-economic status across generations. For instance, growing up in a better educated environment is expected to be positively correlated with human capital formation. Indeed, better educated households are more likely to be better off and thus to invest more in their children's education. Almost two thirds of SHARELIFE respondents declare that there were not enough books to fill one shelf (at most 25 books) in their accommodation at the age of 10. However, there are wide cross-country differences showing that Sweden, Denmark, Switzerland and the Czech Republic present the highest levels of book provision, while Mediterranean countries and Poland show the lowest (for more details, see Chap. 3 of this volume).

This descriptive analysis shows that there is cross-country heterogeneity in human capital accumulation. Remarkably, Scandinavian countries exhibit high participation to financial markets, low entry-age and high levels of human capital. On the contrary, Mediterranean countries are characterized by a lower propensity towards risky investments, higher entry-age in financial markets and lower levels of human capital indicators.

The next section will investigate the relationship between investment behaviour and human capital accumulation in a multivariate framework. On the one hand, we want to show whether cross-country heterogeneity in investment behaviour remains sizeable when controlling for individual characteristics. On the other hand, we will ascertain whether human capital indicators are significantly related to the decision of investing in risky assets if considered jointly and when allowing for cross-country differences in the institutional framework.

4.4 Multivariate Duration Analysis

We run a duration analysis to investigate how the timing of the first investment in risky assets is related to individual and household characteristics. Duration models have the advantage of combining the information coming from both individuals who actually invested in risky assets and those who have not entered financial markets yet. The alternative strategy, of focusing attention solely on the elderly who have had risky assets, may suffer from endogenous sample selection problems, particularly for Mediterranean countries, where financial market participation is low, as shown in Fig. 4.1. However, all of our analysis is conditional upon survival: to the extent that the poor die younger this may lead to biased estimates even in the context of duration analysis models.

The set of explanatory variables used in our regression includes a variety of individual characteristics: country of residence, gender, year of birth, years of education, self-assessments of mathematical skills at the age of 10, number of books in the accommodation at the age of 10 and current monthly household income. In addition, we use previous waves of SHARE data to take into account respondents' answers to four numeracy questions available in the questionnaire. As explained in the previous section, such numeracy indicators complement the self-evaluations of mathematical skills at the age of 10 since they are collected when respondents are 50+ and allow having two distinct measures of ability in mathematics at different stages of the life-cycle.

Our results show that cross-country differentials in the access to financial markets are statistically significant even after conditioning on further individual characteristics related to investment behaviour. Ceteris paribus, Swedes and Danes invest in risky assets earlier. On the contrary, respondents in Italy, Spain and Greece tend to delay their first risky investment. Hence, at a given age, the probability of having never invested in risky financial markets is lower in Scandinavian countries and higher in Mediterranean countries. We use our estimates to calculate the probability of having never invested in risky assets and show how it varies with age. Figure 4.7 summarizes our results. For simplicity sake we show the results only for Sweden, France, Austria and Italy because they summarize the spatial patterns in Europe and in particular the North–South gradient in access to financial markets. For each year of age (reported on the x-axis) this figure reports the probability of having never invested in risky financial markets up to that time. Overall, the probability of having never held risky assets decreases with age in all countries. However, we notice that it is highest for Italy and lowest for Sweden at any age. France and Austria lie in the middle, but the former is much closer to Sweden than the latter.

The number of years of education is an important predictor of the timing of the first risky investment. Those who remain in full-time education longer anticipate the decision of entering financial markets. This result is in line with the hypothesis that education helps understanding the opportunities offered by risky assets via a positive impact on financial literacy. It should be kept in mind that we included monthly household income among the explanatory variables precisely to explicitly take into account that education might additionally affect investment decisions via an income effect. As displayed in Fig. 4.8, at any age, the higher the number of years of education, the lower is the probability of having never invested in risky assets. The widest gap is found comparing those who spent at most 6 years in full time education with those who remained in full time education longer.

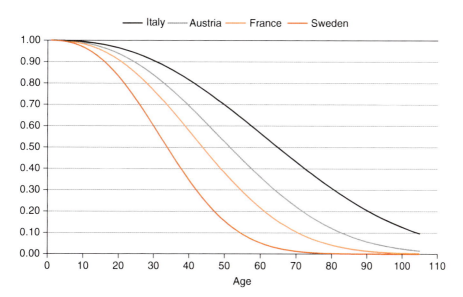

Fig. 4.7 Probability of having never invested in risky assets by age for Sweden, France, Austria and Italy, controlling for birth cohort, gender, education, income, numeracy, mathematical skills and number of books at the age of 10

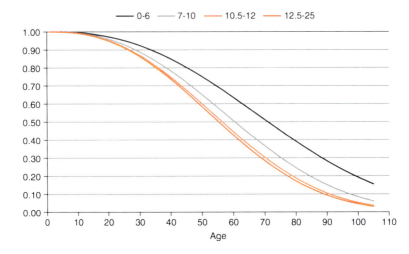

Fig. 4.8 Probability of having never invested in risky assets by years of education, controlling for income, numeracy, mathematical skills, birth-cohort, gender, country effects and number of books in the accommodation at the age of 10

Moreover, individuals with higher mathematical and numeracy skills invest in risky assets earlier. Both indicators consistently show that those who have better numeracy skills anticipate their entry in financial markets. This pattern highlights the crucial role played by mathematics in promoting early financial markets participation, possibly by enhancing the necessary self-confidence in one's ability to make the correct choices. Although our analysis is mainly descriptive and the estimation of causal effects is far beyond its scope, our results suggest that improving the access to education and fostering the teaching of mathematics at all levels of compulsory schooling is a possible strategy to make individuals more familiar with financial markets.

We also find that the cultural background plays an important role in determining investment behaviour. As discussed above, we describe cultural background characterizing respondents' childhood by exploiting the SHARELIFE questions about the number of books in the accommodation at the age of 10. Respondents spending their childhood in an accommodation with more books are more prone to enter financial markets earlier. Number of books is an indicator of the educational attainments of household members. Given the strong and positive correlation between education and the probability of holding risky assets, it is not surprising that respondents growing up in a better educated and economically dynamic environment anticipate their entry in the financial markets. Indeed, their parents, or other older co-habiting relatives, are themselves more likely to hold risky assets, which may make it easier to gather all the necessary information to understand financial markets and decide to invest in risky assets. Finally, we find that women tend to invest later in life, particularly so when they have low education.

4 Human Capital Accumulation and Investment Behaviour

To evaluate the reliability of our empirical results, we calculate for each country the median fitted age of entry into risky financial markets and cross this country-level variable with the ratio between stock market capitalization and GDP, which is a widely used index of financial development (Guiso et al. 2004). Results are reported in Fig. 4.9 (in the case where the predicted value exceeded 100, namely Greece, it was set to 100. Dropping Greece from the analysis makes little difference).

Figure 4.9 shows that there is a clear negative correlation between the age of first risky investment estimated by our model and the development of financial institutions. Although this evidence cannot be conclusive about the causality of this relationship, individuals living in countries where financial markets are less developed delay the decision of investing in risky assets. This is in line with the hypothesis that fostering human capital accumulation might increase the demand for risky financial instruments and contribute to the development of financial markets.

The fitted median age of first investment is even more strongly correlated with the ratio between life insurance premium volume and GDP. The higher this ratio, the higher is the importance of this type of asset in the economy. If we look at Fig. 4.10, we notice that the countries where life insurances are more widespread are also those where the age at first risky investment is lower. In some other countries, notably Italy, Spain and Greece, later entry in financial markets is associated with lower life insurance premium volume. Indeed, in these countries, participation in all financial markets, including life insurance markets, is lower, as documented by Fig. 4.2.

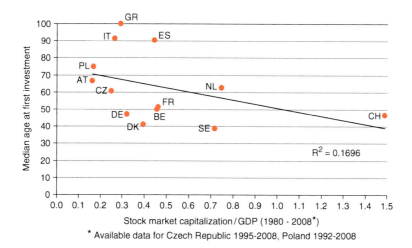

Fig. 4.9 Median predicted age of first entry into financial markets versus stock market capitalization/GDP

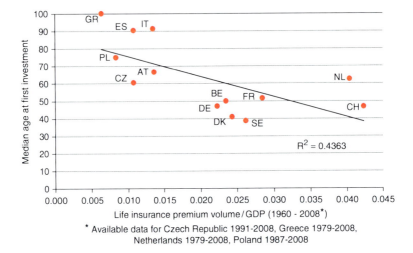

Fig. 4.10 Median predicted age of first entry into financial markets versus life insurance premium volume/GDP

4.5 Conclusions

Making correct use of financial instruments plays an important role in securing a good standard of living in old age. People who failed to invest in equity over the last century missed out on spectacular returns, and are now more likely to be in financial hardship.

Our findings that formal education, mathematical skills and family background all play a role in explaining participation to financial markets comforts the assumption that financial behaviour is affected by levels of human capital. The evidence that in some countries both education and financial market participation is lower for women is of particular interest, given the momentous increase in educational attainment by young females over the last few decades. All this suggests that greater financial literacy constitutes an achievable aim for welfare systems.

References

Angelini, V., Brugiavini, A., & Weber, G. (2009). Ageing and unused capacity in Europe: Is there an early retirement trap? *Economic Policy, 59*, 463–508.

Christelis, D., Jappelli, T., & Padula, M. (2010). Cognitive abilities and portfolio choice. *European Economic Review, 54*(1), 18–38.

Garrouste, C. (2010). *100 Years of educational reforms in Europe: A contextual database*. Luxembourg: Publications Office of the European Union. EUR 24487 EN.

Guiso, L., & Jappelli, T. (2008). *Financial literacy and portfolio diversification*. CSEF Working Paper, No. 212

Guiso, L., Jappelli, T., Padula, M., & Pagano, M. (2004). Financial Market Integration and Economic Growth in the EU. *Economic Policy, 19*(40), 523–577.

Hira, T. K., & Loibl, C. (2008). Gender differences in investment behavior. In J. J. Xiao (Ed.), *Handbook of consumer finance research* (Vol. III, pp. 253–270). New York: Springer.

Lusardi, A. (2008). *Household saving behavior: The role of literacy, information and financial education programs*. NBER Working Paper, No. 13824.

Lusardi, A., & Mitchell, O. (2007). Baby boomer retirement security: The roles of planning, financial literacy, and housing wealth. *Journal of Monetary Economics, 54*, 205–224.

McArdle, J., Smith, J., & Willis, R. (2009). Cognitive and economic outcomes in the health and retirement survey. NBER Working Paper, No. 15266.

OECD (2010). A family affair: Intergenerational social mobility across OECD countries. In *Economic policy reforms: going for growth* (Part II, Chap. 5). Paris: OECD Publications

Chapter 5
The Impact of Childhood Health and Cognition on Portfolio Choice

Dimitris Christelis, Loretti Dobrescu, and Alberto Motta

5.1 Why Childhood Conditions Can Affect Risk Taking in Older Age

Childhood health is by now recognized to influence future educational and economic outcomes. Children who experience poorer childhood health have significantly lower educational attainment, poorer health, and lower socioeconomic status as adults (Case et al. 2005; Currie 2009). For example, Case and Paxson (2008) investigated the relationship between height (as indicator for early health and socio-economic status), cognitive functions and health status at older ages and found that taller individuals (considered to be healthier and wealthier during childhood) have greater cognitive skills on average, report significantly fewer difficulties with activities of daily living, and are in considerably better physical and mental health.

Furthermore, cognitive skills in childhood have their own implications for adulthood. Bad test scores (which could indicate worse cognitive skills) are associated with lower socioeconomic status in childhood, and gaps in abilities by socioeconomic status open up early and tend to persist later in life (Heckman et al. 2006). As captured by achievement test scores measured during education in childhood and adolescence, early cognition is, together with educational attainment, an important determinant of adult socioeconomic success.

Among the many outcomes in adulthood that can be influenced by childhood health and cognition, we will focus on portfolio choice and risk attitudes in the older segment of the population. This analysis is particularly relevant in light of the

---

D. Christelis (✉)
Department of Economics "Marco Fanno", University of Padua, Via del Santo, 33, 35123 Padua (PD), Italy
e-mail: dimitris.christelis@gmail.com

L. Dobrescu, and A. Motta
University of New South Wales (UNSW), School of Economics, Australian School of Business, Sydney 2052, NSW, Australia
e-mail: dobrescu@unsw.edu.au; motta@unsw.edu.au

financial decisions that older individuals face nowadays. Recent reforms of social security systems around the world have been putting individuals more and more in charge of their own financial security after retirement, while the supply of complex financial products has increased. As a consequence, portfolio self-management is becoming commonplace.

Recent literature on the relation between portfolio choice and health (Rosen and Wu 2004) has found that households in poor health tend to hold safer financial portfolios, even after controlling for several other predictors of portfolio choice. In addition, poor health status is associated not only with safer portfolios but also with less diversified ones. The reason is that health shocks in retirement can alter one's financial circumstances and planning horizon, with a direct impact on household economic decisions. When thinking about the relationship between childhood and adult health previously discussed, it might thus well be that adult health is partly mediated by childhood health. We also know that cognition in older age is positively associated with risky portfolio choice (Christelis et al. 2010), and thus, in the same vein, it could also be partly mediated by childhood cognition.

In order to better understand the relationship between childhood health and cognition and risky portfolio choice later in life, we will use the SHARELIFE dataset that provides rich information on childhood health and cognition, and the subsequent financial history of the older population in fifteen European countries, using a harmonized questionnaire that makes results comparable across countries.

5.2 Indicators of Childhood Health and Cognition

The SHARELIFE questionnaire has a rich set of questions pertaining to health during childhood that elicits information on access to care, adverse health events, severe illnesses, vaccinations, hospitalizations, as well as a question on self-reported health during childhood. For our analysis we use, first, the question on whether respondents had access to a usual source of care, that is to a particular person or a place that they went to when they were sick or needed advice about their health. A negative answer to this question could indicate that respondents did not get either the necessary treatment when they became sick or preventive advice needed to forestall health problems in the future. In addition, this lack of medical help could have made them feel more insecure and thus more cautious, which in turn could have shaped their attitude to risk later in life.

Obviously, access to medical care is an important target of welfare and social policies, and a lasting impact of its unavailability in childhood would make such a target even more important to achieve. The prevalence of the availability of a usual source of care during childhood is shown in the first column for each country in Fig. 5.1, and we note that while it is above 95% for most countries, it is lowest for Greece (83%), the Czech Republic (87%), and Germany (90%).

5 The Impact of Childhood Health and Cognition on Portfolio Choice

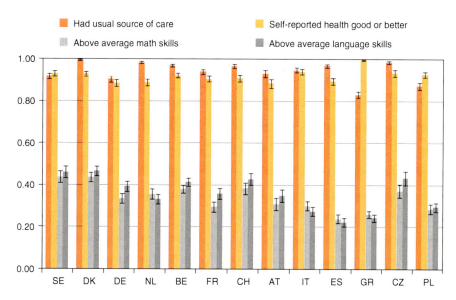

Fig. 5.1 Prevalence of childhood health and cognition variables
Source: SHARELIFE. The number of observations differs slightly by variable, but it is on average approximately equal to 27,650

The second indicator of childhood health that we consider is the subjective health status during childhood as reported by respondents today. This question should provide a broad overview of health status and give an indication of the actual extent of the health problems faced that is not captured by objective measures (e.g. days spent in a hospital). On the other hand, answers to this question could be affected by differences in reporting style across individuals and countries (Jürges 2007). SHARELIFE respondents could rate their health as excellent, very good, good, fair and poor, and we construct an indicator for the highest three answers, the distribution of which is shown across countries in the respective second columns in Fig. 5.1. Prevalence ranges from 88% to 94%, with Austria and Germany being at the lower end of the range, and Italy and the Czech Republic at the high one. The one exception is Greece, in which less than 1% or respondents declare being in fair or poor health during childhood.

Turning now to cognition in early life, we use the answers to two questions that inquire about the respondents' relative position to their peers in school with respect to mathematics and language skills. To this purpose, we construct two indicators for having above average skills. The distribution of those two indicators is shown in the third and fourth country columns in Fig. 5.1, and we note that the range goes from roughly 22% to 45% for both indicators, with Greece and Spain exhibiting the lowest prevalence of above average skills and Sweden and Denmark the highest. As is the case with the self-reported health question, answers to the two childhood cognition questions could be affected by differences in respondents' reporting styles.

5.3 Asset Ownership and Risk Preferences

We now examine the effect of childhood health and cognition on portfolio choice and risk taking in older age. We will concentrate our discussion on four economic choices, namely whether one has ever owned stocks, mutual funds or a private business (this information can be found in SHARELIFE), and whether one is willing to take any financial risks (the latter information is provided in the second wave of SHARE). All three investment choices entail risk taking, and make demands on the cognitive skills of their owners. Furthermore, the assumption of financial risk depends on other sources of risks the investor faces, including health risk. It also depends on the amount of information the investor has on the properties of the different assets, and this in turn is affected by the cognitive skills that the investor has, which were partly formed during childhood.

For each economic choice/health (or cognition) variable combination, we use a probit model in order to associate the choice to the health or cognition variable of interest. We include country fixed effects in order to account for specific conditions prevailing in each country, and we also interact the health or cognition variable with the country dummies in order to allow as much as possible for differential effects across countries. Our variable specification also includes the respondents' gender, education and marital status, and, crucially, indicators of the respondents' socio-economic status during childhood. This is achieved by including in our model variables denoting the number of rooms in the house where respondents lived (adjusted for family size), and the number of books that the respondents' family possessed. As a result, any effects of the childhood health or cognition variables that we find will be net of all the aforementioned factors, and thus less likely to be spurious.

Before discussing our results, it is worth noting that our sample could suffer from survivor bias, i.e. we might predominantly have people in our sample that were physically and cognitively in good shape during childhood, which in turn led to a longer life. To the extent that this is true, the range of values of the health and cognition variables could be somewhat limited and skewed to the higher end of the distribution, and this lack of variation could weaken our empirical results.

The results for stock ownership are shown in Fig. 5.2. We find that the existence of a source of care during childhood is positively associated with stock ownership by roughly 16% points (pp) in Switzerland, 8 pp in Austria, and 5 pp in Italy. Good or better self-reported health matters only in Sweden (9 pp). On the other hand superior mathematics skills boost stock ownership in all countries in our sample except Denmark by roughly 2–11 pp, while language skills show a slightly weaker but still very relevant positive association that ranges from 2 to 12 pp everywhere but Sweden, Denmark, Germany, and Austria. These results are consistent with previous findings that point out the considerable effect of cognitive skills on stockholding.

We now turn to investment in mutual funds, shown in Fig. 5.3. We find that having a usual source of care has a strong positive association with such an

5 The Impact of Childhood Health and Cognition on Portfolio Choice

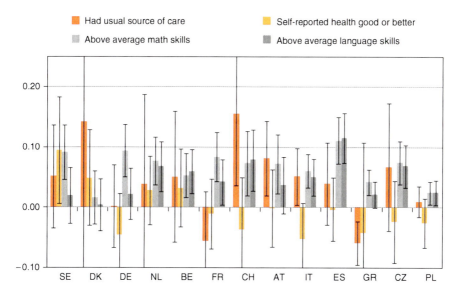

Fig. 5.2 Coefficients (marginal effects) on likelihood of having ever owned stocks
Source: SHARELIFE. The number of observations is approximately equal to 24,850 on average, depending on the specification

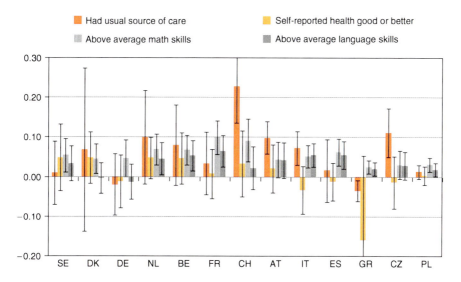

Fig. 5.3 Coefficients (marginal effects) on likelihood of having ever owned mutual funds
Source: SHARELIFE. The number of observations is approximately equal to 24,850 on average, depending on the specification

investment in Switzerland, Austria, Italy, and the Czech Republic, with magnitudes ranging from 7 to 23 pp. On the other hand, good health does not seem to matter anywhere. Above average mathematics skills are once more pretty strong for all countries except Austria and the Czech Republic, with the associations ranging from roughly 2 to 10 pp. Language skills seem to also play an important role for mutual fund investment in the Netherlands, Belgium, France, Italy, Spain, Greece and Poland. We note that the associations for mutual funds are slightly weaker than those for stocks, which is consistent with the fact that the latter are more demanding than the former in terms of performance monitoring and evaluation.

The ownership of a business is to a large extent affected by intangibles like ambition, entrepreneurial spirit, and perseverance, and thus might be less affected by adverse health and cognition status than the previous two assets. Furthermore, circumstances might be such that owning a business is essentially the only career path available (e.g. in agriculture, or when one is shut out from civil service due to discrimination), which means that it does not involve as much risk taking as normally assumed. Perhaps because of these considerations, we find (results are shown in Fig. 5.4) a sizeable positive association with the availability of a usual source of care only in France (3 pp), but we should note that there are no business owners that had fair or poor childhood health in Denmark, the Netherlands and Switzerland, and this lack of sample variability makes estimation for these three countries impossible. Similarly, we find no significant effects of good self-reported health (there are no business owners that had bad childhood health in Greece). Above average cognitive skills are now much weaker than in the case of stocks and mutual funds, with mathematics skills still mattering, however, in Italy (2 pp), Spain (3 pp) and the Czech Republic (1 pp), while language skills don't seem to play any role in ever owning a business.

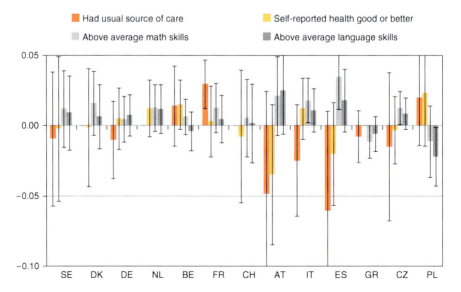

Fig. 5.4 Coefficients (marginal effects) on likelihood of having ever owned a business
Source: SHARELIFE. The number of observations is approximately equal to 24,850 on average, depending on the specification

5 The Impact of Childhood Health and Cognition on Portfolio Choice

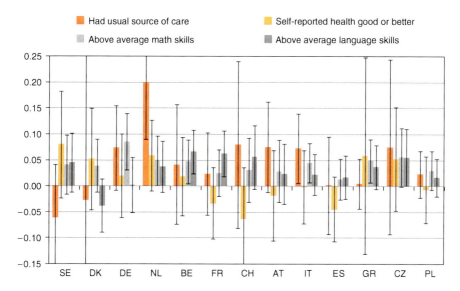

Fig. 5.5 Coefficients (marginal effects) on likelihood of willing to assume some financial risk
Source: SHARELIFE and SHARE wave 2. The number of observations is approximately equal to 16,400 on average, depending on the specification

We would expect that the health risk should negatively affect financial risk taking, as individuals subject to the former might want to control overall risk in their lives by limiting the latter. To that effect, we find results are shown in Fig. 5.5 taking in the Netherlands (20 pp), and Italy (7 pp). On the other hand, the association with self-reported health is not significant anywhere. In addition, we find that above average mathematics skills are strongly positively associated with financial risk taking in Germany, the Netherlands, Belgium, Italy and Greece (magnitudes range from 4 to 8 pp). The importance of language skills is slightly less widespread, but the associations are still relevant in the Belgium (7 pp), France (6 pp), and the Czech Republic (5 pp).

5.4 Summary

Childhood health and cognition have lasting effects on health, cognition and socioeconomic status later in life, and thus directly or indirectly a number of economic choices and attitudes. Lack of a usual source of care is associated with less investment in stocks, mutual funds, while the association with private business ownership and risk taking is quite weaker. The effects of poor childhood health as reported by respondents are weaker and less widespread. Childhood cognition, as represented by above average mathematics and language skills, is strongly positively associated with investment in stocks and mutual funds, and less so with ownership of a private business. The positive association is also strong with financial risk taking.

Our results point to the necessity and economic benefits of welfare policy interventions early in life that aim to increase access to health care providers. Such interventions should also aim to make health care more affordable, and to include initiatives for disease prevention and education on health issues. Furthermore, as our results are not geographically concentrated in any particular area of the European Union, the scope for such interventions seems to be pretty wide.

Given that we find strong evidence that childhood cognition has a lasting impact in many economic outcomes in older age, the potential benefits of interventions aiming at boosting childhood cognitive skills are very high. Such interventions could include more resources for high quality childcare, for remedial education of students that lag behind their peers, for programs that aim to prevent early school leaving, and for post-secondary education training. The cost of such resources should be amply justified by the substantial societal benefits of widespread higher childhood cognitive skills, namely higher human capital formation, decreased income and wealth inequality in adulthood, and an associated increase in social cohesion.

References

Case, A., Fertig, A., & Paxson, C. (2005). The lasting impact of childhood health and circumstance. *Journal of Health Economics, 24*(2), 365–389.

Case, A., & Paxson, C. (2008). Height, health and cognitive function at older ages. *American Economic Review Papers and Proceedings, 98*(2), 463–467.

Christelis, D., Jappelli, T., & Padula, M. (2010). Cognitive abitilies and portfolio choice. *European Economic Review, 54*(1), 18–38.

Currie, J. (2009). Healthy, wealthy, and wise: Socioeconomic status, poor health in childhood, and human capital development. *Journal of Economic Literature, 47*(1), 87–122.

Heckman, J. J., Stixrud, J., & Urzua, S. (2006). The effects of cognitive and noncognitive abilities on labor market outcomes and social behavior. *Journal of Labor Economics, 24*(3), 411–482.

Jürges, H. (2007). True health vs. response styles: Exploring cross-country differences in self-reported health. *Health Economics, 16*(2), 163–178.

Rosen, H., & Wu, S. (2004). Portfolio choice and health status. *Journal of Financial Economics, 72*(3), 457–484.

Chapter 6
Nest Leaving in Europe

Viola Angelini, Anne Laferrère, and Giacomo Pasini

6.1 A North–South Gradient

The nest leaving period and the age at which individuals establish their own independent household are of primary policy concern since they are critically linked to many economic and social outcomes. The choices made by young adults are numerous: further education, marriage, parenthood, first job. All are interrelated and can be linked to another choice, that of a first independent home. Youth labour supply and educational choices will determine the length of the career, pension and life-time consumption. Billari and Tabellini (2008) show that Italians who leave the parental home earlier in life earn a higher income in their mid 30s. This might be due either to the fact that they tend to have longer working experience or to a negative impact of prolonged co-residence on ambitions and motivations for children who leave late (Alessie et al. 2006). The demographic transition and population evolution are largely linked to the timing of first parenthood. Health in later life and life expectancy are linked to the education level.

Previous studies have shown large cross-country differences in the age at which residential independency is established. In 2004 the proportion of men aged 25–29 co-residing with their parents was below 25% in France, the Netherlands, the UK and Australia, while it was above 60% in Mediterranean countries (73% in Italy) and Finland (Becker et al. 2010; Cobb-Clark 2008). We take advantage of

---

V. Angelini (✉)
Department of Economics, Econometrics and Finance, University of Groningen, Nettelbosje 2, 9747 AE Groningen, The Netherlands
e-mail: v.angelini@rug.nl

A. Laferrère
Insee, 18 boulevard A. Pinard, 75675 Paris cedex 14, France
e-mail: anne.laferrere@insee.fr

G. Pasini
Dipartimento di Scienze Economiche, Università Ca' Foscari Venice, Cannaregio, 873, 30121 Venice, Italy
e-mail: giacomo.pasini@unive.it

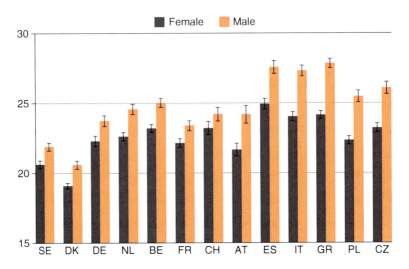

Fig. 6.1 Nest leaving age by country and gender

SHARELIFE to relate the nest-leaving age to life-history. We also use the former waves of SHARE to document the nest-leaving age of the respondents' own children, which helps complement the picture with the behaviour of younger cohorts.

The Europeans aged 50 or more established their own home at a mean age of 24, 90% had left by the age of 30. The current cross country differences already existed in the past: those coming from Southern European countries left their parents' home on average between 2 and 5 years later than their Northern European counterparts (Fig. 6.1). While the average nest leaving age is above age 25 in Spain, Italy and Greece, it is around 20 in Denmark and Sweden. The differences among other countries are smaller. The spread between Scandinavian and Mediterranean countries may reflect cultural differences such as attitudes towards family relationship and youth freedom and independence, linked to ideology and religion, but may reflect also differences in welfare state generosity, availability of housing, of jobs or of higher education. Even if policies and government interventions are largely driven by culture, geography and history, we aim at finding to what extent policies drove living arrangements over the past century.

6.2 The "Push–Pull Effect" of Family

At the family level, altruistic parents help their young adult children in two ways: either by providing their own home for co-residence, or, if unconstrained, helping to pay for another accommodation. Very constrained parents cannot even keep their children at home. Hence there is a non-monotonic relationship between parents' resources and nest-leaving. The quality of the home also matters. If the parents have both low income and a low quality house, the two effects add up and the child has to leave. As the home gets more comfortable, the child can and is induced to stay longer. Finally, if the parents are not resource-constrained, they can help their child to establish its own household by making a monetary transfer, even if at the same time the quality of the home induces the child to stay longer (Laferrère 2005a, b). These "push–pull effects" take place within various national traditions, economic contexts and social policies, which also contribute to pushing young adults out of their parents' home or to keeping them in. As families compare the costs of co-residence and independence, housing costs are likely to be of primary importance (Börsch-Supan 1986; Ermisch 1999; Laferrère and le Blanc 2004). Other elements of context are important: higher education supply, its cost and the geography of this supply; military service, the job market situation and even contraception, which influence partnership behaviour. Entering into all those features is beyond the scope of this chapter and we concentrate on housing and education policies and on education outcomes.

We surmise that the existence of rental accommodation is crucial for nest-leaving, since this phase of life is a period of successive choices (education, partner, job) that often go along with a residential mobility. Homeownership is associated with high mobility costs that make it inadequate for mobile young adults. Rental accommodation was and still is rare or even absent in Greece, Spain, and Italy. In all other countries, with the exception of Belgium (that has a rather high mean age at nest leaving) the rental sector is developed, so the "rental supply explanation" is likely to be important even if only part of the story.

6.3 Life-Time Parent-Child Co-residence

To document how nest leaving behaviour evolved over the last century, we build three broad cohorts: cohort 1 of those born before 1935 (24% of the sample), cohort 2 of those born between 1935 and 1944 (30%) and cohort 3 of those born between 1945 and 1956 (46%), whom we call the "baby-boomers". First we are interested in those who never established their own household and did not give a date for starting to live on their own or establishing their own household. At the other extreme some give a very early date for starting to live "on their own" and they may have come back at their parents later on. We take the answer as it is given. On average 2.5% of all individuals aged 50 or more said they never established their own household. The overall rate is constant over time, but the differences between countries are significant (Fig. 6.2).

In Austria, the percentage of respondents who never established their own household is around 7%; it is around 5% in Spain, 3% in Germany, Switzerland and Italy, and 2% in Greece and Poland. In all other countries, never leaving the parents' home is rare and almost absent, except in the older cohort in France, where it can be seen as a last remnant of the importance of agriculture. In a multivariate

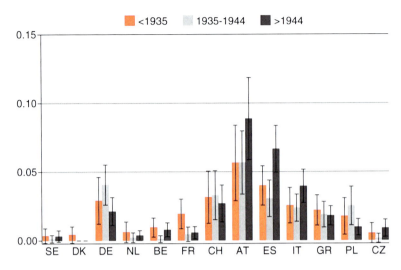

Fig. 6.2 Fraction of respondents who never established their own household by cohort and country

analysis, we proxy socio-economic background by the occupation of the main breadwinner when the respondent was 10; we use the house and family characteristics at age 10 to proxy for home comfort and privacy at the time of nest-leaving, and the location of the parents' home (at the time of departure for those who left and at the survey date for those who never left) to proxy for house prices and education or employment opportunities.

Our results show that never leaving the parents' home is indeed correlated to having parents in agriculture in Germany and France. In that case not establishing one's own household is clearly linked to the life occupation choice. It is correlated to living in a rural area in Austria. In all countries never leaving is more likely for those who remained single, as marrying usually goes with establishing a household. The home characteristics play a role. The child is more likely to stay in the parental home if it provides more space per person. Life-time co-residence with the parents is also more likely for males than for females, a gender difference that will be found for those who left the nest. We leave aside those who never left their parents' home and turn to those who did establish their own household at some point in the life cycle.

6.4 A Historical Decline, Which is Stopped or Reversed for Those Born After the Mid 1960s

Figure 6.3 shows a general decreasing pattern over time: younger cohorts tend to leave the nest earlier. The profile is flat in Denmark, where the mean age was already quite low for the older cohort (20 years). This makes Denmark a special

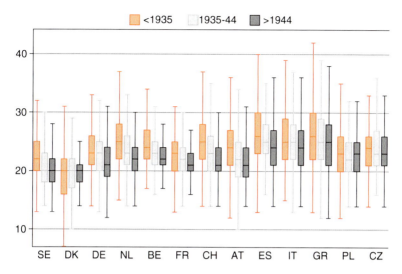

Fig. 6.3 Nest leaving age by country and cohort

case where the independence of young people seems to have always been valued. The dispersion of the nest-leaving ages is higher for the older cohorts, and overall we observe a reduction of the differences both within and between countries over the three quarters of a century under review. The spread (between Spain/Greece and Denmark) went from 8 years for the older cohort, to 5 years (between Spain/Italy and Denmark) for the baby-boomers.

Some information on more recent trends can be gathered from what the SHARE respondents tell about the age at which their own children moved from the parental household (Fig. 6.4). For the cohort 1945–1954 (the baby boomers), the mean age of nest leaving in the children sample as declared by the parents is close to what is directly observed from the sample of respondents. Interestingly, for the younger cohorts the decline in nest leaving age has stopped and also reversed in the Southern European countries, in France and in Belgium. The dispersion in ages also seems to increase in some countries.

6.5 The Influence of Parental Background and Home

We analyze the nest leaving age in two steps: first we look at individual determinants based on what we know of parental and home characteristics; then we turn to contextual policy variables.

The influence of parental background is striking. Even if it varies by country and gender, the overall pattern is clear. Leaving aside children of farmers, male children of both rich parents (professionals or senior managers) and poor parents (in an elementary occupation) left earlier than "middle class" children of blue collar and

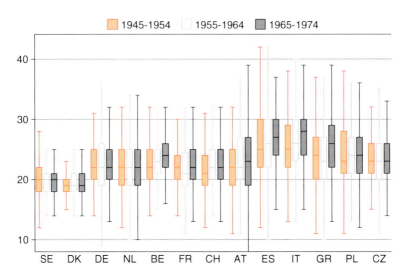

Fig. 6.4 Nest leaving age of respondents' own children, by country and cohort

craft workers. In most countries children left earlier if there was no bread winner in the household at age 10. They were more likely to stay longer if the parental home was comfortable or offered more rooms per person, ceteris paribus. Living with a step-parent, or in a three generation family also induces to move earlier, pointing to privacy reasons. Even with our very crude proxy for parental income, the non-monotonic relation between nest leaving and parental background confirms the theoretical predictions of a push–pull effect: some well-off parents can afford helping their children to move out, while poor families are unable to keep them (especially daughters) at home. Once the home characteristics are taken into account, the relationship between our proxy for parental resources and nest leaving age is hump-shaped for daughters, but flatter for sons, pointing to the important pushing out effect of the various home characteristics, and underlining the fact that children of richer parents can afford to leave earlier.

That housing prices play a role is vindicated by the important influence of the location of the parental home at the time of nest leaving. Ceteris paribus, a child living in a large city, where housing is likely to be more expensive leaves half a year later than one in a small town; a child in a village or rural area, where housing is cheap, leaves earlier. The price effect may be mixed with the fact that children living in rural areas or villages leave because they have to move to town to find a job or study. Indeed making the move from a rural area to a city advance nest-leaving by more than year, ceteris paribus. In Poland, Italy or Greece, the children who leave earlier tend to be those of farmers or of non-executives, pointing to some children being constrained to move out. Those who left a parental home situated in a foreign country also left younger.

The multivariate analysis confirms the important cohort effect: compared to the baby-boomers our eldest cohort left some 2.2 years later, the middle cohort 1 year later. Two factors have concurred to this tendency. First, age at marriage, which is positively related to the age of nest leaving, has, in most SHARE countries, *decreased* by one year over the period, contributing to around half a year decrease in the overall decline in nest leaving age. Second, the age at which young people left education, which is negatively related to the age of nest leaving (once age at marriage is controlled for; more on this below), has *increased* over cohorts by 3.1 years, contributing to between a quarter and a half of a year in the overall decline in nest leaving age. Modifications in marriage and education behaviour thus "explain" 36% of the decline in nest leaving age. Other factors must have played a role. As income has increased over the century we interpret this unexplained earlier nest leaving as a relaxing of the elements constraining youth independence in the past.

6.6 The Importance of Housing Policies

The unexplained cohort evolution and our findings about the importance of the parental background naturally lead us to verify that the age at which the parental nest is left is linked to the socio-economic context of the period. To test it we rely on country and time specific context variables, related to housing and education policies at the time our respondents reached adulthood. Needless to say that such contextual variables may also capture unobserved country and time effects, hence the results should be taken with caution; they are nevertheless striking (Fig. 6.5). The oldest cohort reached adulthood during the war or just after it, at a time of acute housing shortage in most countries; besides, the hard rent control that was present in all countries at that time was detrimental to young outsiders and delayed nest-leaving. In the 1950s and 1960s the construction of subsidized rental housing (either a

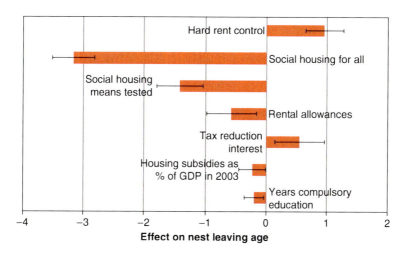

Fig. 6.5 The effect of housing policies on nest leaving age

universal right in Sweden, Denmark and the Netherlands or means-tested in Germany, Belgium, France, Switzerland and Austria), and later the introduction of rental allowances, helped to reduce the nest leaving age. Coupled with the underdevelopment of rental accommodation in the Southern countries and Belgium, the high nest leaving age could also be accounted for by the absence of housing credit markets (Alessie et al. 2006). However, for those cohorts, we do not see a direct influence of credit liberalization or homeownership policies on the nest-leaving age. Further investigation is clearly required. Another possible explanation is the role of family ties in Catholic countries such as Italy, Austria and Spain (Reher 1998). This could reinforce the negative effect on nest leaving age of the underdevelopment of mortgage and rental markets, and at the same time explain some characteristics of the welfare state designed during the second half of the last century. For example, in Italy there are almost no unemployment benefits, nor publicly provided long term care institutions for the elderly. The burden of financial and care needs of the unemployed and the disabled are implicitly assigned to the families: therefore, co-residence can act as a mean to alleviate those costs.

As for education policy, which we capture by the number of years of compulsory education that has increased over the years its effects is that of a small *decrease* in nest leaving age. Even after controlling for the socio-economic determinants, the order of countries in nest leaving age is not altered, as Denmark is still where children leave earlier than in all other countries, and the Mediterranean countries where they leave the latest, followed by Poland, the Czech Republic and Belgium. Once controls are introduced the differences between countries are not reduced over cohorts, which points to deep cultural differences.

The recent increase in the nest leaving age is not analysed here. It may be linked to youth unemployment and show a contrario how low unemployment of the late 1960s induced the early nest leaving of the baby-boomers; it also might be linked to the fact that more young people pursue higher education (partly in order to avoid unemployment), and do so at their parents' home, more than in the past because homes are more comfortable and the supply of higher education has increased and spread. However, housing supply and the recent increase in rents and house prices in some places are likely to play a primary role.

6.7 Women Leave Earlier than Men, but Leaving Directly to Marry Has Declined

In all SHARE countries women leave one or two years earlier than men (Fig. 6.1). This partly reflects differences in the age of marriage. On average women of those cohorts married 3 years earlier than men. In most countries women are more likely than men to establish their own household directly by getting married (Fig. 6.6). It is interesting to notice the cross country distribution of marriage-motivated nest leaving: Eastern and Southern Europeans, together with Belgians, tended to leave

6 Nest Leaving in Europe 75

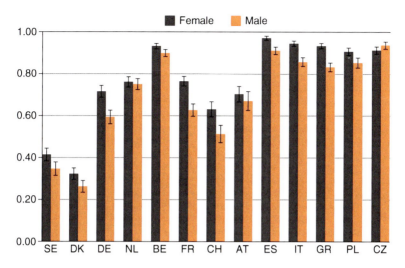

Fig. 6.6 Nest leaving with marriage by country and gender

their nest only when they got married. Belgium has little renting accommodation, as the Southern countries. On the contrary, less than half of the Scandinavians left their parents' home in order to marry. This may be due to differences in accessibility to rental housing or differences in marriage rates, with partnership being more developed there. When we include cohabitation, the percentage of respondents who left the nest to live with a partner or marry increases, in Sweden and Denmark more so than in the other countries. However, the cross-country pattern remains the same.

The two panels of Fig. 6.7 split the sample by cohort and gender. In general, the percentage of people leaving the nest to get married declined over time, and was much higher for the older cohorts. In spite of the decline in age at marriage, gradually for the younger cohorts and first for males it became more common to live by oneself before marriage. The cohort evolution is particularly strong in Sweden and Denmark, where the fraction of people leaving the nest to marry halved between the oldest and the youngest cohort. Similar declining patterns can be observed for some other countries (France, Germany, Switzerland, Sweden, Denmark, the Netherlands, Austria), although at various levels. The evolution was much slower in Belgium, Greece, Poland, Czech Republic, and has hardly begun in Italy and Spain. In spite of this common trend, the three country-grouping pattern (Belgium, Spain, Italy, Greece, Poland and Czech Republic; Sweden and Denmark; Germany, the Netherlands, France, Switzerland and Austria) is stable over time.

Multivariate analysis (not shown here) shows that indeed one is more likely to leave to directly get married in countries where no rental accommodation is available for young single persons. The existence of social housing and of rental allowances decreases the likelihood to leave directly to marry, while hard rent control or the tax deduction of mortgage interest, a policy supposed to encourage home ownership, increase it. To summarize bluntly: rental accommodation is for

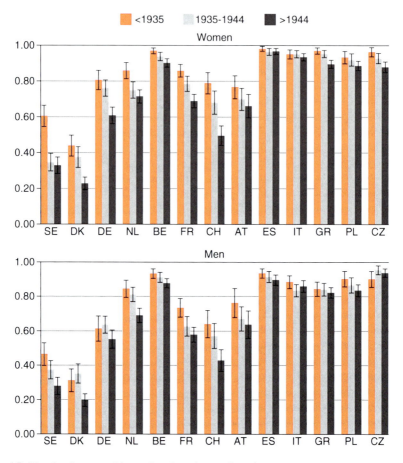

Fig. 6.7 Nest leaving age with marriage by cohort and gender

singles, home ownership is for couples. The likelihood to leave directly for marriage is higher for children of poorer background (breadwinner in elementary, blue collar or craft occupations). It is also negatively related to education (the more years of education the less "direct marriages"). Living on one's own before getting married, forfeiting economies of scale in accommodation appears to have been a luxury.

6.8 Nest Leaving and Education: Leaving Early is Good!

The link between marriage and education choice, and the fact that education has important consequences in terms of health, life expectancy and well being of the elderly population, both for men and women, suggest to study more closely the link between nest-leaving and education.

When plotting the nest leaving age against years of education, we identify three groups of countries (Fig. 6.8). Spain, Italy and Greece feature late departure and low education (and as we said above, low welfare state, small rental sector); Sweden, Denmark are characterized by early departure, high education (and high welfare, large rental supply); the other countries lie in between.

It is interesting to note that the negative correlation seems stronger for the baby boomers than for the older cohorts. Differences across generations reflect the spectacular development of education over time. Indeed the mean age at which they finished education was 14.3 for the older cohort and 17.4 for the baby-boomers. The percent that pursued education after age 18, which might be even more relevant, went from 17.7 to 35.8% and 24.8% for the middle cohort.

Splitting the sample by gender, we see that women lagged somewhat behind: half as many females as males were into education above age 18 in the older cohort

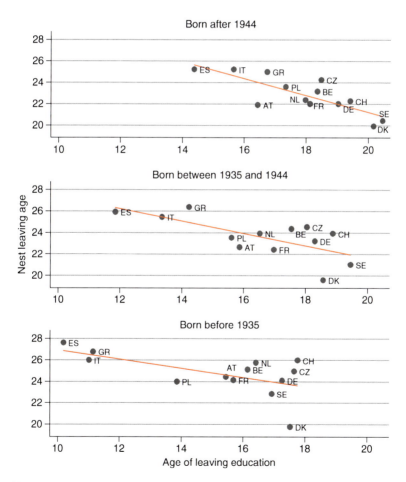

Fig. 6.8 Nest leaving age and age of leaving education by cohort

(13.5 vs. 24.6%), although the difference has narrowed for the baby-boomers (32.4 vs. 39.4%). The links between higher education and nest leaving behaviour is not straightforward as education might have been provided close to the parents' home. Then an important quality of the parents' home is its location. A large majority (85%) of the SHARE respondents spent all their student years still living with their parents: from 90% or more in the southern countries, in Belgium, Poland, Austria and the Czech Republic, to less than 50% in Denmark. This again seems linked to the availability of rental housing that helps leaving the parents to study. The higher the education level, the less likely one is to get it at the parents' home, although urban youths are more likely to have got it nearby. Children of richer parents are more likely to leave them to study, and the tendency to leave has increased over cohort, ceteris paribus. Note that females are also less likely to study at home than males, which points to women being more independent than men of the same age and education level. If women leave their parents earlier it is not only because they marry, but also because their education made them more able to cope with living independently.

We try to confirm the macro level negative relationship between nest leaving age and age of leaving education at the micro level to document whether the nest leaving age per se has really long lasting consequences in terms of well-being. As many factors (such as supply and location of higher education) are interrelated, pinpointing the channels of the effect is difficult. The intuition is that in countries where a significant proportion leave their parents to study the education level of the SHARE respondents will be negatively linked to the age at which they left. The results are summarized in Fig. 6.9. At first glance, there seems to be a *positive* correlation between nest leaving age and age of leaving education: the longer you stay home, the longer you study, and the later you start your first job. However, the correlation becomes negative after controlling for marriage behaviour: you studied

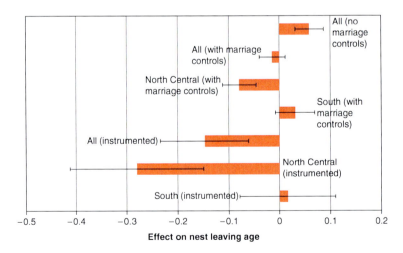

Fig. 6.9 Nest leaving age and age of leaving education: micro-level analysis

longer if you left home earlier. More precisely it is negative in all but the Southern countries where the correlation remains positive. In Spain and Italy less than 3% from those cohorts left their parents to pursue further education, so the absence of correlation is not surprising. Leaving aside the three Southern countries, the negative correlation is even more pronounced when the age of leaving education is instrumented by maths and verbal abilities of the respondent at age 10 (which we assume influence the length of education but not directly the nest leaving behaviour; we also use number of books and number of years of compulsory education as instruments). Overall, the microeconomic analysis confirms the macro relationship between age at nest leaving and higher education. Being able to leave the nest enhanced the chances to pursue further education ceteris paribus. This seemed particularly true in France, Germany and the Netherlands.

6.9 Conclusion

The nest leaving period is linked to crucial life choices: higher education, marriage, first job. The age at which young adults establish their own household is deeply country specific: it ranges from age 20 in Denmark to around 25 in the Southern countries. From our study of the individual and the macroeconomic determinants of nest-leaving age we draw the following main conclusions:

- Nest-leaving age is shaped by individual parental and home characteristics, but is also linked to the national and policy context. The age of compulsory education and the availability of rental housing influence the nest-leaving age. Over the last century we documented a decline of the age at marriage, a decrease of the frequency of leaving to directly marry, an increase in higher education and in the frequency of getting this education after having left the parental nest. All of these have concurred to earlier nest-leaving. If at first approximation the economic situation is exogenous, the differences between countries suggest that the housing market, and especially the availability of affordable first rental homes, plays a primary role. All in all, more favourable economic conditions allowed leaving earlier.
- On average, 78% of the SHARE respondents left and married directly, more so for women than for men. The historical decline of this practice leaves more time for higher education and parallels the increase in education level.
- We also looked more deeply into the link between education and the age of nest-leaving. Ceteris paribus, it seems that an earlier nest-leaving is positively related to education level. The higher the education the more you get it by living outside your parent's home.
- The declining trend in nest-leaving age has spectacularly stopped, or even reversed for the more recent cohorts who tend to stay home longer. It might be linked to deteriorating economic conditions in some countries for those born in the 1970s, even if other factors might play a role. Some may pursue higher

education in order to avoid unemployment, and do so at their parents' home, more than in the past. However the SHARELIFE data strongly suggests that providing first rental accommodation for the young enhance the chances of higher education.

References

Alessie, R., Brugiavini, A., & Weber, G. (2006). Saving and cohabitation: the economic consequences of living with one's parents in Italy and The Netherlands. In R. H. Clarida, J. A. Frankel, F. Giavazzi, & K. D. West (Eds.), *NBER international seminar on macroeconomics 2004* (pp. 413–441). Cambridge, MA: MIT Press.

Becker, S., Bentolila, S., Fernandes, A., & Ichino, A. (2010). Youth emancipation and perceived job insecurity of parents and children. *Journal of Population Economics, 23*, 1047–1071.

Billari, F., & Tabellini, G. (2008). Italians are late: does it matter? In J. B. Shoven (Ed.), *Demography and the economy, NBER Books*. Chicago: University of Chicago Press.

Börsch-Supan, A. (1986). Household formation, housing prices, and public policy impacts. *Journal of Public Economics, 30*, 145–164.

Cobb-Clark, D. A. (2008). Leaving home: What economics has to say about the living arrangements of young Australians. *Australian Economic Review, 41*, 160–176.

Ermisch, J. (1999). Prices, parents, and young people's household formation. *Journal of Urban Economics, 45*, 47–71.

Laferrère, A. (2005a). Leaving the nest: the interaction of parental income and family environment. INSEE WP 2005-01.

Laferrère, A. (2005b). Quitter le nid: entre forces centripètes et centrifuges. *Économie et Statistique, 381-382*, 147–175.

Laferrère, A., & le Blanc, D. (2004). Gone with the windfall: How do housing allowances affect student co-residence? *CESIFO Economic Studies, 50*, 451–477.

Reher, D. S. (1998). Family ties in Western Europe: Persistent contrasts. *Population and Development Review, 24*, 203–234.

Chapter 7
Homeownership in Old Age at the Crossroad Between Personal and National Histories

Viola Angelini, Anne Laferrère, and Guglielmo Weber

7.1 Homeownership as Old Age Insurance

Compared to other forms of savings for old age, homeownership offers some advantages. Purchasing a home is similar to purchasing an annuity that would insure housing consumption. Moreover, the home may be seen as a secure asset in case of need and perceived as a substitute for the purchase of long term care insurance. It is also a family asset that may be transmitted to the next generation. These advantages are weighted against the drawbacks of over consumption in old age (for those who are house rich and cash poor), low portfolio diversification (the price risk may be important if all assets are in the home), and illiquidity (drawing equity in case of need is not easy).

A majority of the European elderly are homeowners. Homeownership rate is above 70% for those aged 50–79. The life cycle model of saving under borrowing constraints predicts a hump shaped homeownership age profile (Artle and Varaiya 1978). The ownership rate increases with age as people save and become homeowners, and declines in old age as people draw on their housing equity. The model is actually complicated by large mobility costs and housing illiquidity. This delays homeownership to a time when mobility is expected to be lower, and reduces equity withdrawal. Hence the decline in old age of home ownership is open to debate.

V. Angelini (✉)
Department of Economics, Econometrics and Finance, University of Groningen, Nettelbosje 2, 9747 AE Groningen, The Netherlands
e-mail: v.angelini@rug.nl

A. Laferrère
Insee; 18 boulevard A. Pinard, 75675 Paris, cedex 14, France
e-mail: anne.laferrere@insee.fr

G. Weber
Dipartimento di Scienze Economiche "Marco Fanno", Università di Padova, Via del Santo 33, 35123 Padova, Italy
e-mail: guglielmo.weber@unipd.it

Homeownership age profiles and rates vary within a country among groups of population (by income, say); across countries and over time for different cohorts. This variation might stem from differences in national housing policies, such as credit constraints and mortgage market laws, rental regulations and social housing availability. It might also stem from personal housing history, such as parents' homeownership, inheritance, or migration. Personal histories are themselves shaped by more general history and country policies, just as housing policy is in turn influenced by the local political and economic context.

7.2 A Digest of Housing Policies in Europe

Housing has been a target for government intervention in most countries, both in terms of welfare (because a home is a necessity; because of housing consumption externalities), and in terms of credit market development (housing is an expensive investment). Besides, the last century saw two world wars that heavily destroyed homes or halted construction, the Great Depression before WWII, and, after WWII, a baby boom that mechanically increased home space demand, then overall home demand after 1965 when the first baby boomers began setting up their own households. Housing shortage was a familiar feature of the 1950s until the end of the 1970s. Those characteristics make housing policy both a social policy and an economic policy. Housing conditions were seen as linked to health and welfare, hence part of a desirable "social" policy, but in most countries the housing stock remained private property and the market played some role. Besides, new construction fuelled growth, hence the broader economic policy aspect. The aims of government intervention were always somewhat mixed, and the form it took has varied in space and time.

Indeed, many forms of public intervention (at various levels, local or national) have existed over the last century. Nominal rents were blocked during or after WWI in Belgium, Switzerland, Sweden, France, Germany, Italy, Austria, Spain (1939), after WWII in Denmark and Greece. Construction was subsidized through various forms of "social" rented dwellings, mostly after WWII (Sweden, Denmark, Germany, France, Austria and, then, the Netherlands and Italy). After 1970, direct allowances for rents began to be granted to low-income tenants, a departure from the indirect subsidy of social housing that was more universal as in most countries exceeding the income eligibility requirement was not a reason for leaving social housing or paying higher rent (France, Germany) and in some places entry into subsidized housing was not even means-tested (Sweden, Denmark and the Netherlands). Even after the hard first-generation rent controls were dropped, a variety of second generation rent regulations and sitting tenants protection are still prevalent in most if not all continental European countries.

After the period of direct subsidy to social housing, but sometimes in parallel, home ownership was encouraged. The classical arguments in favor of homeownership are the following. Homeownership reduces eviction risk, rent risk and avoids the "fundamental rental externality" of Henderson and Ioannides (1983).

Eviction risk has been close to zero for most SHARE cohorts, as tenants were well protected. Rent risk is quite low for most sitting tenants, even if it may become important in the near future. The fundamental rental externality whereby tenants take little care of their home, inducing a moral hazard premium in market rents, may explain why so many consumers prefer owning to renting and why it might be in the public interest to promote it. Since WWII, homeownership has been encouraged by subsidized interest rates; by tax incentives such as deduction of mortgage interest from taxable income (Sweden, Denmark, the Netherlands and France); no taxation of imputed rent (except in Belgium, Sweden, Denmark, the Netherlands, Spain, Italy and Greece); no taxation of capital gains on homes; or even, lately, the sale of social housing to tenants (the Netherlands, Sweden, Italy and the former Eastern Block countries). For low-income, low-tax people the advantages of tax breaks are limited. Also, tax incentives may be fully capitalized in house prices, making first time buying more difficult. For this reason, there was a move towards deregulation of mortgage markets in all countries: after the 1970s in Germany, at the end of the 1980s in Austria, France, Spain, Sweden, in the 1990s in Greece, Italy, Poland and the Czech Republic. It has been shown that the development of the mortgage market helped young people's access to homeownership (e.g. via increased loan to value ratios).

We use SHARELIFE data also to document the two main movements that drove the housing life cycle of Europeans: the move to urban areas and renting, then the move towards more suburban areas and ownership. While 43% of the childhood homes of those born before 1935 were in a village or a rural area, it is the case for only 36% of the baby boomers. The drive led to a huge increase in comfort. When those born before 1935 were aged 10, 46% had no running water, nor toilet, nor central heating in their home; only 21% of the baby-boomers lived in similar conditions in their childhood. However, most subsidized rental housing was built in the 1960s with more emphasis on quantity than quality, and most if not all were flats. The reason for the wide spread move toward homeownership may then also be found in the desire for "different" homes (typically not available for rent because of the externality described above): more space, houses and better neighbourhoods, as opposed to flats in city centers or suburbs. Moreover, there might have been a filtering up process through social housing as a step towards home ownership. The intuition is that in countries with a well-developed social housing sector (Denmark, Sweden and France) access to ownership was also easier ceteris paribus: it left time to build up a down payment, to settle in a job and a family, and helped young people move out (see Chap. 6).

7.3 Homeownership Rate by Age, Cohort and Country

SHARELIFE offers the unique opportunity to envision a century of access to homeownership in Europe. It allows getting at "true" age profiles free from cohort effects, at least for those who survived to the day of the interview. The well-documented differential mortality by wealth implies that we may overestimate the proportion of homeowners in the overall population. The presence of time effects may also blur the picture.

We define three broad cohorts: cohort 1 of those born before 1935 (24% of the sample), cohort 2 of those born between 1935 and 1944 (31%) and cohort 3 of those born between 1945 and up to 1954 (45%), whom we call the "baby-boomers". Even more important than date of birth is the date around which each cohort started to set up their own household: before the end of WW2 for cohort 1, after WW2 for cohort 2 and at the end of the 1970s for cohort 3.

In Table 7.1 we show the proportion of individuals who are or have ever been home-owners (after they left their parents' home), by cohort and country. This is an underestimate of the proportion of individuals who experience home-ownership at any point in their lives, as some individuals may become home-owners during their remaining life span. This underestimation may be of some consequence for cohort 3 – at least in some countries, where mortgage markets function less well and owning through inheritance is more common.

There are marked differences between cohorts: barely 65% of our eldest cohort ever owned their home, the rate jumps to 68% for the middle cohort and for the baby boomers. The difference between cohort 1 and cohort 2 is spectacular in the Nordic countries, in the Netherlands, Germany, Switzerland or France where it is between 6 and 11% points and corresponds to the development of the credit market for homes and reconstruction after WWII. Those born after 1934 (cohort 2) started benefiting from it at the time they formed a household, after 1944. There are some exceptions: in Greece the rate is stable at a high 86%. The evolution for the baby-boomers (cohort 3) is less spectacular, and in some countries (Sweden, France, Spain, Italy, Austria) they were even less likely than the preceding cohort to ever own a home.

Figure 7.1 presents the complete age profile of homeownership for the three cohorts, by country. We define ownership as 1 if the person (or her spouse) owned during the entire 10-year age period, 0 if she never owned, and y/10 if she owned for y years during the 10-year period. One caveat has to be mentioned: the samples get small for countries with low homeownership rates (Austria, Germany, Poland and Czech Republic).

We are now able to qualify the cohort effects in more detail. The curves show (cohort-specific) age profiles. The vertical distance between curves measures a first type of cohort effect: the difference between cohorts in homeownership rates at each age. Another type of cohort effect is the variation from one cohort to the next in the slopes of the curves, which measure the steepness of the age profiles.

Table 7.1 Percentage of individuals who ever owned a home, by cohort and country

Cohort	SE	DK	DE	NL	BE	FR	CH	AT	ES	IT	GR	PL	CZ	All
3. >1944	84.9	93.2	55.0	80.5	90.3	77.6	62.0	61.5	74.5	67.4	87.3	57.2	61.5	68.4
2. 1935–1944	87.2	90.0	54.8	67.8	86.7	79.7	62.2	65.1	77.0	72.9	85.4	54.7	56.6	67.9
1. <1935	80.7	83.1	48.5	57.0	83.5	72.7	52.9	61.2	74.6	69.4	86.3	51.9	51.2	64.7
Total	84.5	90.0	53.4	71.9	87.5	76.7	59.7	62.6	75.2	69.6	86.5	55.5	58.1	67.3

7 Homeownership in Old Age at the Crossroad

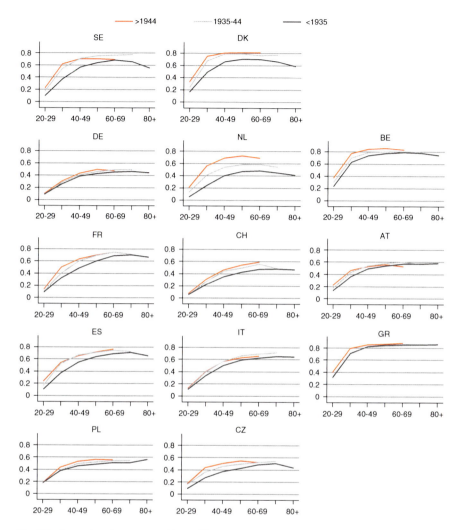

Fig. 7.1 Age-cohort homeownership profiles

7.3.1 Age Profiles

The age profiles look the same nearly everywhere: an increase in homeownership up to age 50–59, then a leveling up and a small moving out of ownership in old age, after 70 in Denmark, Sweden and the Netherlands, and rather after age 80 in the other countries (except Poland and Greece where no moving out of homeownership is apparent).

This pattern confirms the life-cycle theory that predicts equity withdrawal in old age as a way to maintain a stable standard of living. To the extent that financial instruments to release home equity are not available, selling and moving into rented accommodation may be an efficient way to free resources. The alternative is to draw equity by buying cheaper homes, as documented in Chap. 8.

7.3.2 Cohort Effects of the First Type: The Spread of Home Ownership

Let us look at the first type of cohort effect, measured by the vertical distance between the curves. If one curve is above another, it means the corresponding cohort has higher ownership rates than the other at all ages. At first glance, such cohort effect is large between cohort 1 and cohort 2, in Sweden, Denmark, France, Spain, the Netherlands, (West) Germany (not shown separately on Fig. 7.1) and the Czech Republic; it is smaller in Belgium, Austria, Switzerland, Poland and Italy. This effect could be due to public policies encouraging home ownership in the 1960s and 1970s. Besides home prices were low and real interest rates were very low because of inflation. By contrast, in most countries, the progression towards homeownership stopped for cohort 3, with the exception of the Netherlands and East Germany (not shown separately in Fig. 7.1). While some will still enter homeownership later in life, the slower progress toward homeownership might come as a surprise as it coincides with the progressive liberalization of credit. It could be that mortgage development was eaten up by increases in home prices, which hit first time homebuyers, together with the economic crisis of 1973–1979.

7.3.3 Second Type of Cohort Effects: Earlier Access to Home Ownership Thanks to Credit

Even if the profiles look alike, they are shaped by different histories. Such cohort effects of the second type can appear by a look at the steepness of the age profile. For all cohorts, homeownership is infrequent below 30, except in Greece where more than 25% own. In Greece a third of these young home owners report they received it as a gift or bequest. The proportion of heirs of this young age group is also high in Spain, Italy, Austria, Poland and Czech Republic, but the ownership rate is lower (less than 20%). The slope between the first two age groups (20–29 and 30–39) is steepest in Greece and Belgium. In those two countries the profile is rather flat above 40. However, the story differs: as we said, inheritance is prevalent in Greece, where help from family is also frequently mentioned (21%); only 6% mention a gift or inheritance in Belgium, where 76% of the 20–29 and 75% of the 29–39 got a mortgage. The interesting point is that the slope gets steeper and steeper from one cohort to the next in many countries. We interpret it as following the development of mortgage and credit markets which made it easier to borrow against future income or to access mortgages at a younger age in those countries.

7 Homeownership in Old Age at the Crossroad

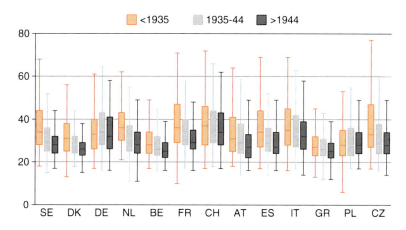

Fig. 7.2 Distribution of the age of first home-ownership, by country and cohort

Indeed, Fig. 7.2 shows that from one cohort to the next not only access to home ownership increased, but it happened earlier. Median first home ownership age was 31 for the owners of cohort 1, 29 for cohort 2, and 27 for cohort 3. Among those who occupied a home as owners, 31% (58) purchased or got one before age 30 (age 40) for the older cohort, 36% (67) for the middle cohort, and 46% (75) for the baby boomers.

These findings point to very large differences between cohorts in homeownership profile. In the Netherlands 48% of the oldest cohort owned when aged 60–69, while for the baby-boomers currently aged 60–69 the home ownership rate is 72%. In France the rate went from 67 to 76% and in Denmark from 72 to 82. Even if the increase was smaller in some other countries home ownership rates of the future elderly are high. The question of using this housing equity optimally to finance old age has to be raised.

7.4 Tenure and Location Over the Life-Cycle

A home is a place to live, and owning and renting differ in that respect, as most owned homes are houses in low density areas, when rentals are flats in high density areas. Over the life cycle the most likely move is from the parental house in a rural area to a flat in the city, to a house in the suburbs or a more rural area, and finally sometimes to a flat in a city again. This can clearly be observed for our cohort 1: 33% of their parental homes were owned and in rural areas or small town, 31% of the independent homes where they moved before age 30 were rented in cities or suburbs, 42 (56)% of homes where they moved in between age 30 and 39 (between age 40 and 59) were owned, then 50% of homes where they moved in between age 60 and 69, down to 16% after 80 as rent free becomes the main tenure mode in case of a move.

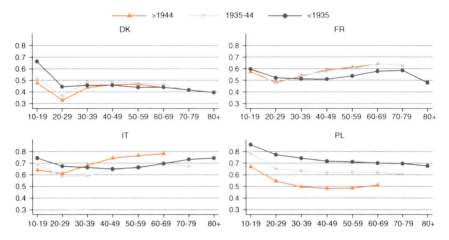

Fig. 7.3 Age-cohort profiles of living in a rural area or small town

Figure 7.3 presents the age profiles of the rate at which individuals live in a rural area, a village or a small town, for our three cohorts, for four example countries. This includes home-owners, renters, but also those living rent-free (in the parental home when young, in their children's home when old) or even in nursing homes.

In all countries and for all cohorts there is a dip on entry into adulthood when people move to the city. This is normally followed by a period when the age profiles are flat, then slowly rising as some city flat dwellers move to low density areas as they enter into homeownership and houses. At the very end of life the profiles are flat again, or even declining as the oldest old prefer to be closer to the services provided by cities. The cohort effects vary strikingly by country. In countries such as Poland at the bottom right of Fig. 7.3 (Greece, Spain, and the Czech Republic would be the same), each cohort is less likely to live in a village or small town setting than the preceding one, following the decline in agriculture that took place earlier in the other countries. No return to the cities is seen there, as indeed the homeownership profiles are also rather flat, except for the Spanish younger cohort who has entered into the process. In countries such as Belgium, France, Germany or Austria the adult younger cohort is more likely to live in a low density area than the oldest cohort, as the development of homeownership went with a clear move towards houses and gardens. This seems also to have happened more recently in Italy. In old age the location choices of some baby-boomers might become an issue if services are located in more densely located areas.

7.5 How Mortgages Help to Become Homeowners

Figure 7.4 presents the proportion of owners who accessed homeownership via a credit or a mortgage when they were aged below 40, by cohort and country. The increase between cohort 1 and cohort 2 is more than 20% points in France (from 58

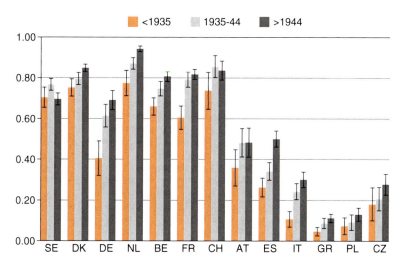

Fig. 7.4 Access to credit for purchase of a home before age 40, by country and cohort

to 79%, catching up with the two Nordic countries, the Netherlands, Switzerland and Belgium where mortgages where already developed before WWII) and Germany (from 42 to 65%), even if this country somewhat lags behind. The increase, from an already high level, has been important in the Netherlands where most if not all of the younger cohort homeowners got a mortgage. For some other countries, housing credit did not benefit a majority of those cohorts, even if the progression over the century has been important: Spain, Italy and the Czech Republic.

As the mortgage industry developed, other means of access to homeownership declined (Fig. 7.5). For our older cohort in "central Europe" countries (Czech Republic, Poland, Austria, Germany), Italy and Greece more than a quarter of those who accessed homeownership before age 40 did it through a family inheritance or a gift. The proportion was only between 10 and 20% in Spain, Switzerland, France and Belgium, and much lower elsewhere. For the baby-boomers the proportion is lower, but it remains strikingly high (around 25%) in Poland, Czech Republic, Greece, Austria and Italy. It is still 16% in Germany.

The decline in inheritance helped democratize access to homeownership. However, the family remains an important source of help for all cohorts. Even if people do not inherit a family home directly, the parents help them buying it (Fig. 7.6). The help is mentioned by more than 10% of baby-boomers (who acquired a home before age 40) in Austria, Greece, the Czech Republic, Switzerland, Germany, Italy and Poland. We find again the opposition between Central- East- South-Europe and North-West.

In Figs. 7.7 and 7.8 we show that there is a striking correlation between the development of the mortgage market and how people access homeownership. In Fig. 7.7 we plot the proportion of households who bought their house through a mortgage before age 40 against a very crude indicator of mortgage market development, that is the maximum loan-to-value ratio on mortgages (LTV) in the early

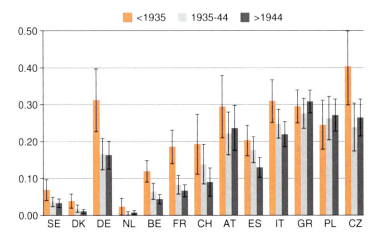

Fig. 7.5 Inheritance and gift as a means of access to homeownership before age 40, by country and cohort

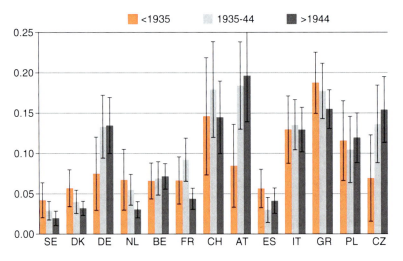

Fig. 7.6 Purchase of a home before age 40 with help from family, by country and cohort

1980s. There is a clear positive relation: where the mortgage market is well developed, more people use it for their home purchases. At the same time, in countries where the mortgage industry is more developed, other means of access to homeownership, such as inheritances, gifts and family help, are less widespread (Fig. 7.8).

These relations hold also over time. In countries where the maximum loan-to-value ratio significantly increased between the early 1960s and the early 1980s, access to formal credit increased more than in countries where it did not change (the percentage difference between the two groups of countries is 8.1%), while resorting to family help, gifts and inheritances became less common (-7.0%).

7 Homeownership in Old Age at the Crossroad

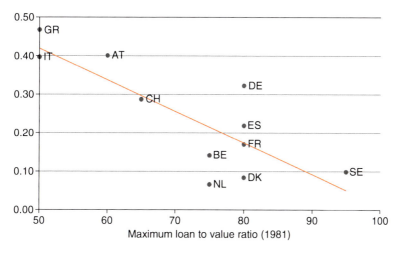

Fig. 7.7 Proportions who bought their home with a mortgage before 40 and maximum loan-to-value ratios

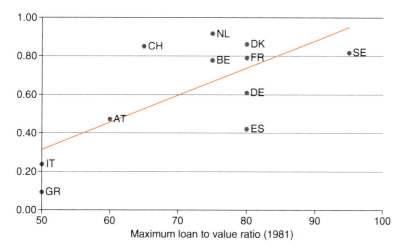

Fig. 7.8 Proportions who received their home as a gift or acquired it with help from family before 40 and maximum loan-to-value ratios

7.6 Conclusions

The secular changes in home-ownership rate and age patterns across European countries and cohorts are at the crossroad between the personal life histories of the SHARE respondents and broader national histories. They are related to housing policies, credit market development and socio-demographic characteristics of each country.

Our main conclusions are as follows:

- The overall lifecycle pattern is largely similar for all cohorts and countries: going from a parental home in a rural area, to being a tenant in a city, to owning (often in a less densely populated area). In some countries, renting is less common, and help from the parents is instrumental in acquiring a home.
- The differences in both location and slopes of home-ownership age profiles across cohorts are strong. Those born after 1935 had a much larger probability to ever be homeowner than those born before. For those born after 1945, the overall probability is not larger but the access to homeownership took place much earlier in their lifecycle.

Our analysis may have implications for policy makers:

- The high homeownership rate of elderly Europeans, coupled with their relatively low levels of financial wealth in some countries, suggests that there is room for introducing new efficient home equity conversion products for those who might need to supplement a low pension.
- The location of some the new cohort of elderly in villages or small towns may call for a reorganization of old age care.

References

Artle, R., & Varaiya, P. (1978). Life cycle consumption and ownership. *Journal of Economic Theory, 18*, 35–58.

Henderson, J. V., & Ioannides, Y. M. (1983). A model of housing tenure choice. *American Economic Review, 73*, 98–111.

Chapter 8
Does Downsizing of Housing Equity Alleviate Financial Distress in Old Age?

Viola Angelini, Agar Brugiavini, and Guglielmo Weber

8.1 Housing and Financial Distress

Reasons for residential moves vary across the life-course. While deteriorating health and increasing needs for support by children have been suggested to be the main motivations for moves in later life, one as yet under-investigated reason might be the wish to reduce home equity, thus alleviating financial distress in old age.

Housing is the most widely held asset and, therefore, an important component of household wealth in many European countries. At the same time, it provides services: by owning their home, consumers can insure against the risk of housing service inflation (Sinai and Souleles 2005). For this reason, it is not surprising to find that very high proportions of elderly households own their home in all SHARE countries. However, if they want to keep a good standard of living, they should release home equity by either taking up a mortgage, or by downsizing, or both.

We know from the first two waves of SHARE, that large fractions of households report difficulties making ends meet. This is particularly true of renters, but is quite common among home-owners as well. In fact, Angelini et al. (2009) argue that the failure to use financial instruments that reduce home equity (like mortgages or reverse mortgages) late in life is partly responsible for financial hardship among elderly Europeans. In this paper we document the extent to which elderly Europeans

---

V. Angelini
Department of Economics, Econometrics and Finance, University of Groningen, Nettelbosje 2, 9747 AE Groningen, The Netherlands
e-mail: v.angelini@rug.nl

A. Brugiavini
Dipartimento di Scienze Economiche, Università Ca' Foscari Venice, Cannaregio, 873, 30121 Venice, Italy
e-mail: brugiavi@unive.it

G. Weber
Dipartimento di Scienze Economiche "Marco Fanno", Università di Padova, Via del Santo 33, 35123 Padova, Italy
e-mail: guglielmo.weber@unipd.it

downsize their home equity position by trading and show that also the downsizing decision is related to the development of mortgage markets.

The development of mortgage markets varies a lot across European countries. In some countries credit rationing policies were used by central banks to fight inflation until a few decades ago, and until recently banking sector competition was severely restricted. In other countries the poor operation of the judicial system has also been blamed for the limited development of mortgage markets. In all countries low levels of financial literacy may also account for reluctance to use debt instruments and failure to use them well.

There is a striking correlation between indicators of mortgage market development and proportions of elderly households reporting financial distress. For instance, in Fig. 8.1 we plot the proportion of households participating in the second wave of the SHARE survey who report finding it difficult or very difficult to make ends meet in 2006–2007 against a very crude indicator of mortgage market development, that is the typical loan-to-value ratio on mortgages over the 2003–2006 period. There is a clear negative relation (the correlation is −0.59), despite the fact that high loan-to-value ratios in countries like Spain and Greece probably do not apply to ageing mortgagors.

Calza et al. (2009), propose a much more sophisticated index (not available for Switzerland, Poland and the Czech Republic) that takes into account not only loan to value ratio, but also opportunities for and costs of equity withdrawal, typical terms and conditions of mortgage contracts, and so on. Figure 8.2 clearly shows that there is a strong, negative relation between mortgage market development and financial distress in old age (the correlation is −0.54). Although this relation might be partly driven by the fact that Sweden, the Netherlands and Denmark are very different from the other countries, it confirms the evidence reported in Fig. 8.1.

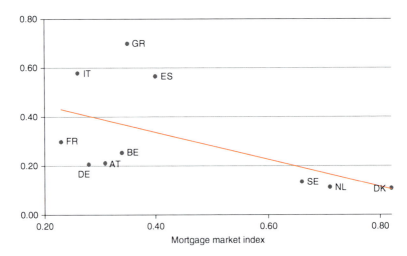

Fig. 8.1 Proportion of homeowners who report difficulties making end meets versus typical loan-to-value ratios by country

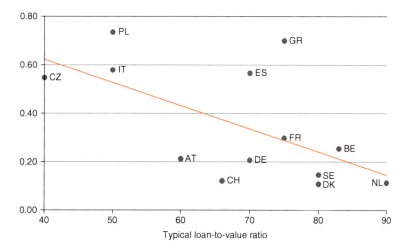

Fig. 8.2 Proportion of homeowners who report difficulties making end meets versus index of mortgage market development by country

Of course, in countries where mortgage refinancing is not available and mortgages are mostly short term, equity withdrawal cannot be achieved by use of debt instruments.

A more radical way to reduce home equity is to sell and move into rented accommodation. This is relatively uncommon, though. Chiuri and Jappelli (2010) use repeated cross section data to show that few households cease to be home owners late in life. In Chap. 7, Angelini, Laferrère and Weber document a similar pattern using life histories data.

We instead document the extent to which elderly Europeans downsize late in life by selling an expensive home and buying a cheaper one. We do this by exploiting a unique feature of the SHARE life history data, which contain individual records of housing transactions over the entire life of the respondents. For each sale or purchase, respondents are asked to report the value of the property, and this allows us to determine whether the respondent was trading down or up.

8.2 How Often Do Europeans Move?

A first indication of housing mobility is given by the total number of homes individuals ever had in their lifetime (so far). This information is available in the data, and can be compared across countries, to the extent that the age distribution is the same.

Figure 8.3 shows that there are major differences across European countries: Northern Europeans report an average number of 6–8 different main residences over their life course, while Southern and Eastern Europeans typically report less

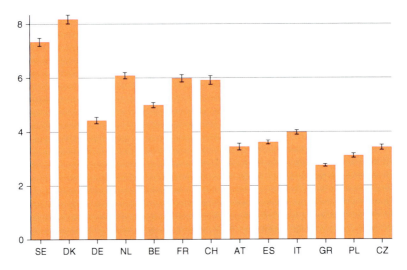

Fig. 8.3 Average number of main residences by country

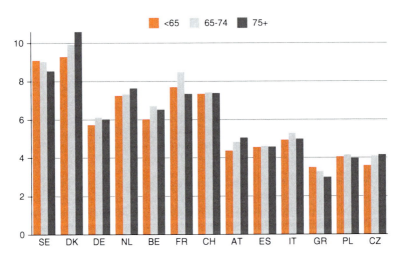

Fig. 8.4 Average number of main residences by cohort and country

than four. Given that this includes the parental home, an average Greek, Polish or Czech apparently moved only at the time of nest-leaving.

Figure 8.4 shows that the average number of main residences hardly increases with age, with the notable exceptions of Denmark and the Netherlands. For instance, younger Greeks and Swedes have already experienced more mobility than their elders. This picture cannot distinguish between age and cohort effects, but if moves in old age were common we would likely observe increasing patterns. Note also that in most countries (with the partial exceptions of the Netherlands, Sweden and Denmark, as far as the first wave of SHARE is concerned), nursing

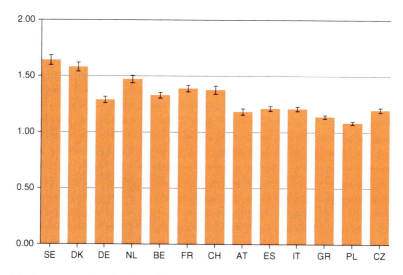

Fig. 8.5 Average number of main residences after age 50, by country

home residents were not included in the baseline sample and this might bias downward the estimated number of moves for the 75+. The extent of the bias is likely to vary across regions – it is well known that nursing homes are more rarely used in Southern European countries.

If we are interested in trading down in old age, it makes sense to look at the number of main residences owned or rented after age 50. We know from previous studies that non-durable consumption peaks around that age (Attanasio et al. 1999), and household size also tends to decrease after that age (with important cross country differences). Figure 8.5 shows that in most countries a majority of individuals never change main residence after they reach age 50: only in Denmark, Sweden and the Netherlands the average number of main residences exceeds 1.5.

8.3 When Do Older Europeans Move?

In Fig. 8.6 we present histograms of ages at which a home purchase or sale took place for four countries where trading decisions are relatively frequent: Sweden, Denmark, the Netherlands and France. The figure shows that a vast majority of transactions takes place relatively early in life, and certainly before retirement age (that we can assume to be around 60, even though it varies a lot across countries, as shown in Angelini et al. 2009).

Figure 8.7 shows the age distribution of trading up decisions, that is, of cases where the respondent bought a more expensive home. Figure 8.8 shows instead for the same countries the age distribution of the more infrequent trading down

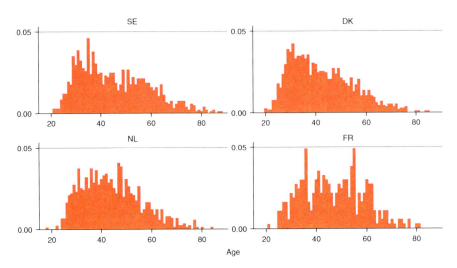

Fig. 8.6 Age distribution of own–own main residence transactions

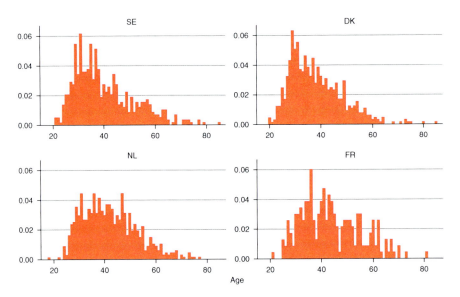

Fig. 8.7 Age distribution of trading up transactions

decisions. Comparing these two figures we see that in all four countries trading down takes place later in life compared to trading up.

Finally, in Fig. 8.9 we provide direct evidence on the frequency of home equity withdrawal in later life by means of moving homes. The graph presents the average number of moves that involve trading up, trading down or going from home-ownership into non-homeownership (labeled as "own–rent" for simplicity sake) that have

8 Does Downsizing of Housing Equity Alleviate Financial Distress in Old Age?

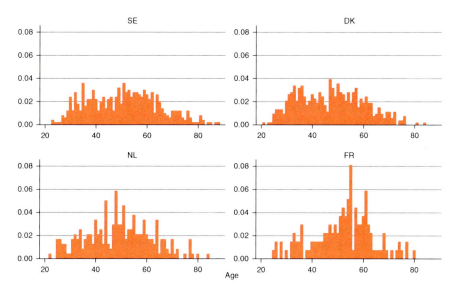

Fig. 8.8 Age distribution of trading down transactions

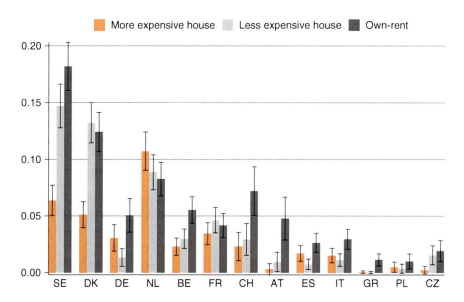

Fig. 8.9 Average number of transactions after age 50, by type and country

taken place after the respondent reached age 50. In all countries, there are more moves that involve some equity release than trading up transactions. We also find that in those countries where mortgage markets are least developed (Southern and Eastern European countries), the number of overall moves are quite low.

8.4 Why Don't Older Europeans Trade Down More?

The striking feature that emerges from our analysis is that the low development of mortgage markets not only limits the ability to withdraw equity by using mortgage debt, but also has a negative correlation with the number of own–own transactions later in life, as shown in Figs. 8.10 and 8.11 (the correlation of the number of own–own transaction with the loan-to-value ratio is 0.66, while it is 0.87 with the mortgage market index).

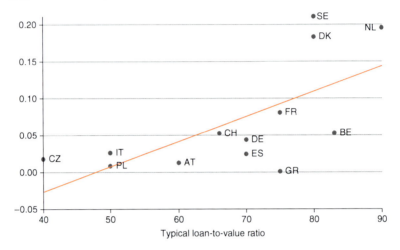

Fig. 8.10 Percentages of own–own main residence transactions after 50 and typical loan-to-value ratios, by country

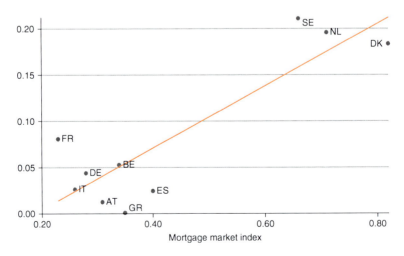

Fig. 8.11 Percentages of own–own main residence transactions after 50 and mortgage market index, by country

In both cases, we show on the vertical axis the ratio of the number of own–own housing transactions carried out after age 50 to the number of homeowners at age 50. On the horizontal axis we report the two indicators of mortgage market development that we already introduced in the introduction, namely the typical loan-to-value ratio for mortgages and the broader mortgage market index constructed by Calza et al. (2009).

8.5 Conclusions

The availability of recall price data on home purchases and sales is a unique feature of the third wave of SHARE. These data can be used to address the question of whether elderly Europeans trade down in old age, as a way to avoid becoming house-rich, cash-poor.

The importance of trading down as a form of equity release depends heavily on financial and mortgage markets access and regulations, as well as on the availability of public housing and long term care accommodation. In most European countries financial instruments that allow equity release are unavailable or relatively uncommon, and cheap public housing is a scarce resource (particularly for the elderly), so trading down may well be the only way to generate a cash flow out of the available home equity. But the evidence we produced suggests that in those countries where mortgage markets are less well developed, lower fractions of home-owners trade down by selling and buying, and higher fractions of homeowners report financial distress.

References

Angelini, V., Brugiavini, A., & Weber, G. (2009). Ageing and unused capacity in Europe: Is there an early retirement trap? *Economic Policy, 59*, 463–508.
Attanasio, O. P., Banks, J., Meghir, C., & Weber, G. (1999). Humps and bumps in lifetime consumption. *Journal of Business and Economic Statistics, 17*(1), 22–35.
Calza, A., Monacelli, T., & Stracca, L. (2009). *Housing finance and monetary policy*. European Central Bank, Working Paper 1069.
Chiuri, M. C., & Jappelli, T. (2010). Do the elderly reduce housing equity? An international comparison. *Journal of Population Economics, 23*, 643–663.
Sinai, T., & Souleles, N. S. (2005). Owner-occupied housing as a hedge against rent risk. *The Quarterly Journal of Economics, 120*, 763–789.

Chapter 9
Separation: Consequences for Wealth in Later Life

Caroline Dewilde, Karel Van den Bosch, and Aaron Van den Heede

9.1 Separation as an Increasingly Important Issue

Over the course of their lives, a substantial minority of elderly European men and women have experienced the dissolution of one or more partner relationships through divorce or the ending of cohabitation. So far, most research into the economic consequences of (marital) separation has been based on panel data and consequently focuses on the short or middle term for the current generations of respondents in their "adult" years. This chapter makes a start at improving existing knowledge by studying the economic consequences of (marital) separation, looking at this issue "from the other way around". Using retrospective SHARELIFE-data, we explore how wealth in later life – measured in terms of home-ownership and the possession of other financial assets – is influenced by the relationship trajectories of European men and women. Given the complexity of the issue at hand, in this chapter we focus on the first marital separation. As for the variation between European countries in terms of the institutional arrangements influencing the short and long-term consequences of (marital) separation, we furthermore sketch the contours of a conceptual framework that can be used for a more in-depth study.

Although in recent years the economic consequences of (marital) separation have been the topic of a fair amount of publications, most authors have concentrated on the short or middle-term impact of this life event. The results of these studies for different European countries are fairly consistent: while many divorced women become financially deprived and/or even end up claiming welfare, for men the financial consequences of divorce are usually less negative – controlling for the

---

C. Dewilde (✉)
Department of Sociology and Anthropology, University of Amsterdam, OZ Achterburgwal 185, 1012 DK Amsterdam, The Netherlands
e-mail: c.l.dewilde@uva.nl

K. Van den Bosch and A. Van den Heede
Herman Deleeck Centre for Social Policy, University of Antwerp, Sint Jacobstraat 2, 2000 Antwerp, Belgium
e-mail: karel.vandenbosch@ua.ac.be; aaron.vandenheede@ua.ac.be

size and the composition of their new household, some men even experience an improvement of their disposable income. These gender differences can be related to the gender division of labour, the gender differences in labour market participation and male–female wage differentials. From a comparative perspective, Uunk (2004) has shown that in countries providing either high public childcare and/or generous transfers for lone parents, the short-term economic consequences of divorce are less severe. Furthermore, not all women follow the same income trajectory during the years following divorce. According to Andreß et al. (2006), British and German women, although they suffer more initially, recover rather quickly from the negative economic consequences of separation, perhaps because their economic situation forces them to repartner more quickly. In Sweden however, where gender equality in terms of a comparatively smaller income loss following separation is most pronounced, both men and women seem to financially suffer for a longer period of time. This might be an indication that in the longer term the incomes of separated women gradually recover, though not always to their previous levels. The mechanisms by which the initial decline in household income following (marital) separation is mainly countered are paid work and/or repartnering. Across Europe, women who repartner experience a 26% increase of their post-divorce income, while women who enter the labour market gain 19% (Dewilde and Uunk 2008).

This focus on (women's) income position has resulted in a somewhat one-sided view of the economic consequences of (marital) separation. In addition to the financial consequences, divorce entails significant additional economic changes. Perhaps the main other economic impact concerns the housing situation of the ex-partners. By definition, the dissolution of the couple results in two separate households, of which at least one has to find another place to live. Setting up a new household comes at a substantial cost, and in particular for those people at the lower end of the income distribution, finding new housing which is both affordable and meets certain quality standards is a difficult task. Likewise, the partner who remains in the marital home often has to provide for the full mortgage or rent payments, which can cause severe financial strain and/or result in moving away. Many respondents, especially women, have difficulties coping with housing costs and maintenance, or find that they do not manage to raise the capital necessary to buy their ex-partner's interest in the house.

Comparative research on the housing consequences of (marital) separation (Dewilde 2008, 2009) shows that in the short-term, this life event substantially raises housing costs, as well as the risk of leaving owner-occupation for both men and women. This risk seems however smaller for women in those countries where the economic consequences of divorce are more severe, i.e. in the liberal and Southern-European welfare regimes. This finding might be explained by the fact that women in these countries might be compensated for the economic consequences of divorce by more advantageous procedures of property settlement. However, other explanations are possible as well. In the Southern-European countries, outright ownership is more common and part of the "extended family enterprise", which makes it easier to remain in owner-occupation. Furthermore, alternative forms of housing are limited, so that divorced men and women might be forced to stay in owner-occupation, even if they cannot afford it.

9.2 In the Long Term

Concerning the long-term consequences of (marital) separation, our first hypothesis is that both men and women who ever experienced the dissolution of a marriage or cohabitation are less likely to be a home-owner and have a lower level of financial assets in old age, even when they have eventually repartnered. This effect is caused by three underlying mechanisms. The first mechanism relates to the event of (marital) separation itself. As the previous section has shown, across Europe (marital) separation results in a decline of disposable household income (especially when taking housing costs into account) for a usually extended period of time, and in the exit out of owner-occupation. We also expect to find a gender difference in the impact of (marital) separation in old age. The second mechanism is related to the well-known beneficial effect of marriage. Wilmoth and Koso (2002) found that marriage, as the institutionalised way of cohabiting, offers more opportunities for welfare accumulation compared to other living arrangements. Thirdly, especially in so-called home-ownership countries, owner-occupation is heavily subsidised through tax credits, benefiting average- and high-income households. Moreover, during the post-war decades, most European countries have experienced a more or less sustained increase in house prices, making owner-occupation a relatively safe and profitable investment. To the extent that renters do not enjoy the same financial benefits, the event of exiting owner-occupation in itself following (marital) separation simply puts separated men and women in a position where they can accumulate less wealth.

9.3 The Importance of Institutions

The impact of divorce obviously depends on the social context. A historical and comparative perspective is therefore important. In this section we single out two dimensions: the welfare state and the legal traditions concerning the division of marital property. Furthermore, as the respondents in the SHARELIFE-module grew up, established their families and bought their homes in different time periods, we have to take into account that processes and mechanisms of wealth accumulation (both housing wealth and other financial wealth), as well as the impact of (marital) separation on these, might be different for different birth cohorts. Therefore, in our analyses we distinguish between respondents born from 1900 to 1934, from 1935 to 1944 and from 1945 onwards.

Concerning the possibly mitigating influence of the welfare state on the impact of (marital) separation on wealth in old age, we expect that female labour market participation can be regarded as a reflection of the opportunity structure for women.

This is influenced by welfare state interventions in matters like public child care and maternity benefits, but also by education and legislation towards non-discrimination at the work place. Of course, cultural and economic factors are also important. In our empirical analysis, we use the female labour market participation rate in 1980 as a kind of "catch-all" contextual variable for this opportunity structure. This choice is mainly motivated by the lack of data on specific welfare state interventions in the earlier decades for all countries. In any case, we expect that women in countries with high female labour market participation are more likely to be economically independent, and therefore suffer less from negative consequences on their wealth after separation.

A second institutional domain we at least want to draw attention to concerns the legal customs related to the division of marital property. So far, this domain has remained a black box, and is not often addressed by sociologists. From a comparative perspective however, laws and legal customs potentially have a large impact on the effect of marital separation on wealth in old age. Issues such as compensation for the "wronged" spouse in terms of facilitating ownership of the family home or the sharing of pension rights have a large impact on economic well-being and the possibilities for wealth accumulation for both separated men and women. Furthermore, the change from bilateral to unilateral divorce laws might have an impact on the economic outcomes in later life remains. These questions however remain unanswered up to today as research is still scant on these issues.

9.4 Data and Variables

The data used in this chapter are based on the marital histories collected in SHARELIFE. These retrospective data are then combined with the prospective data collected during the second wave of SHARE. The main reason for combining the SHARELIFE data with the second wave of SHARE is that the dependent variables we use as indicators for economic well-being were not collected in SHARELIFE. The data refer to the life histories of 28,573 respondents in 13 countries (Austria, Germany, Sweden, The Netherlands, Spain, Italy, France, Denmark, Greece, Switzerland, Belgium, Poland and the Czech Republic). Using listwise deletion, cases with missing information on the variables of interest were removed from the sample, producing an analytical sample that contained a total of 20,711 respondents, including 4,552 currently singles and 16,159 currently married or cohabiting individuals. In the following paragraphs we discuss our main variables and our analysis strategy.

9.4.1 Dependent Variables

Our analyses focus on home-ownership and the total net worth (i.e. wealth) of SHARELIFE-respondents. Home-ownership as recorded in the second wave of

SHARE is measured as a dichotomous variable, coded 0 if a respondent does not own the house he or she occupies at the time of interview and coded 1 if he or she does. The total net worth is the sum of the net values of: (a) the primary residence net of the mortgage, (b) other real estate, (c) business, (d) cars, (e) savings, stocks and bonds, mutual funds, IRA's and life insurances. Imputations on net worth are provided in the SHARE-data to correct for missing values. This net worth is expressed in PPP-adjusted Euros. It is important to note that in SHARE net worth is measured at the household level. This is problematic for our analysis since it implies that net worth is confounded with current marital status. For couples, the total net worth represents the wealth of two adults, whereas for singles it reflects the wealth of one adult. Given our goal of estimating the effect of individuals' marital histories on their economic well-being in later life, we need to adjust for this by creating a per capita measure of total net worth as dependent variable. To do so we assign half of the total household net worth to each partner by dividing the household level wealth by the number of people living in the household. After inspection of the data we applied a logarithmic transformation of our dependent variable, net worth, to correct for the non-normal distribution. To avoid the loss of respondents with negative or zero net worth values, we added a constant to the total wealth distribution to anchor the minimum value at 1 before applying the log transformation. Using the logged net worth changes the interpretation of the continuous coefficients, which can now be interpreted as percentage changes. Dummy variables, however, cannot easily be interpreted as percentage changes. Following Halvorst and Palmquist (1980), the coefficients for the dummy variables are transformed by taking the antilog of the coefficient, subtracting 1 from this antilog and multiplying the result by 100 to obtain percentage changes.

9.4.2 Independent Variables

Since the main goal of this study is to examine the long-term effect of different marital trajectories on economic well-being, we created marital status categories based on the current marital status and the previous marriage dissolutions and remarriages. The two possible current marital statuses are married/cohabiting or single. Non-cohabiting couples were grouped in with singles (236 respondents). We grouped respondents that are married, but living separated from their spouses, in with married respondents (22 respondents). Grouping the currently married or cohabiting individuals according to their marital history results in four categories: (1) the continuously married/cohabiting (reference), (2) individuals who never married, (3) remarried after experiencing at least one divorce, and (4) cohabiting after experiencing at least one divorce. We distinguish between three categories of singles: (5) never married, (6) single after experiencing at least one divorce and (7) single following widowhood. We excluded individuals who experienced a combination of widowhood and divorces (329 respondents) and individuals who remarried or cohabited after the death of a partner because of the small sample numbers

(350 respondents). We also excluded the individuals for whom we could not identify marital history (751 respondents) and those who cohabited but never married (181 respondents). Note that this reconstruction of marital history does take into account neither the sequence nor the number of dissolutions. Another important note is that the results for the groups of cohabiting divorcees and widows and widowers should be interpreted with caution because of the small numbers.

The institutional effects of the different welfare states are estimated using the percentage of female labour-force participation in all countries in 1980, the earliest year for which we have data for all countries (ILO 2010).

9.4.3 Control Variables

In all models we control for age, age squared, birth cohort (1900–1934, 1935–1944, 1945–...), education (Low, Medium, High), subjective health (Very Good to Excellent, Less than Very Good), number of chronic diseases, number of living children, number of siblings, the degree of urbanisation, and the European region one is living in: (1) North (Sweden, Denmark, The Netherlands); (2) West (Austria, Germany, Belgium, France, Switzerland); (3) South (Spain, Italy, Greece); and (4) East (Poland, Czech Republic) as dummy variables.

9.5 Results

Let us first look at the distribution of the sample across marital history categories. Figure 9.1 shows that a large proportion of our sample is still in their first marriage (66% overall). Thirty percent experienced a marital dissolution, while 4% never married or cohabited. The dominant reason for marital dissolution for women born before 1945 and men born before 1935 is widowhood; for later cohorts divorce becomes more important, which reflects the increasing divorce rates since the 1960s. Among men, about half of the divorcees are currently remarried (53%), while 33% is currently single and 13% found a new partner, but did not remarry. Among female divorcees, the proportion that is still single is much larger. About 55% is still single after experiencing a divorce. While the number of women who cohabits after divorce is quite comparable to that of men (10%), the number of remarried women is relatively small (34%).

What is the impact of marital history on the rate of home-ownership? Unsurprisingly, Fig. 9.2 shows that continuous marriage or cohabiting is most conducive to ownership; people who never married are less often owner-occupiers. People who experienced a marital dissolution are also less often owner-occupiers. The interesting result is that among those who eventually remarried or entered cohabitation, the rate of ownership is still about 15% points lower, compared to those who never experienced marital dissolution. Those who were owner-occupier but remained

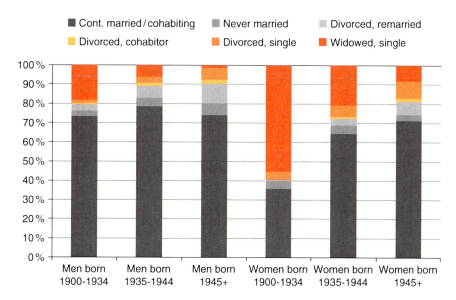

Fig. 9.1 Distribution of marital history by birth cohort and gender

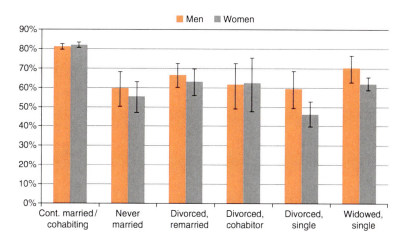

Fig. 9.2 Percentage owner-occupiers, by gender and marital history

single after divorce are less likely to be owner-occupier afterwards. This probably reflects the fact that divorce implies that at least one partner has to move and is thus at risk of exiting home-ownership. In general, there is not much difference between men and women, except for divorcees who are now single, where women appear to fare worse than men.

The patterns are a bit less clear when we look at net wealth (Fig. 9.3). Among men and women, individuals who never married enjoy about the same mean level of net wealth as persons who are still in their first marriage. Divorcees who remarried or

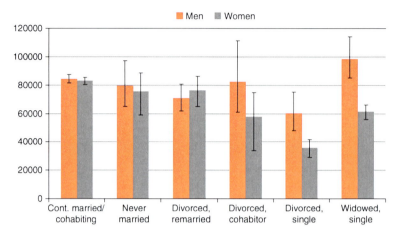

Fig. 9.3 Mean (geometric) net wealth per person in PPP adjusted euro, by gender and marital history

started cohabiting have a slightly lower mean level of net worth, although the difference is not statistically significant. Individuals who remain single after divorce possess low levels of wealth, and this is clearly true for women. Men who were confronted with the death of the partner and remained single have a higher mean net worth than women in the same situation, and have a higher mean net worth than continuously married individuals. Of course, these results could be due to selection effects, or cohort or country differences, which is why we now turn to the regression results.

In the regressions we model the natural log of wealth per capita (implying that the regression coefficients can be interpreted as percentage changes), controlling for age, age squared, birth cohort, education, subjective health, number of chronic diseases, number of living children, number of siblings, the degree of urbanisation, and the European region one is living in. Regressions are performed for men and women separately, and by European region. Given the focus of this chapter, we only plot the coefficients for the marital history dummies in Fig. 9.4a, b. The reference category refers to the people who are continuously married. For the models with home-ownership as dependent variable we use logistic regression, but otherwise the model is the same.

Since confidence intervals are large, relative to the estimated coefficients, we must be careful when drawing conclusions. First, in general, marital dissolution has a negative impact (if any) on net wealth and the chance to be a home-owner. Divorcees – both men and women – who remain single, seem to suffer most. The impact is much reduced, and sometimes wholly eliminated, when divorcees remarry. Widowhood seems less disadvantageous than divorce, probably because it does not involve dividing up the assets between partners. In the Northern countries, the impact of marital dissolution is about the same for men and women. This could be due to the higher labour market participation of women in the Scandinavian countries. In the countries of the West (AT, BE, DE, FR), by contrast, divorce and

9 Separation: Consequences for Wealth in Later Life 111

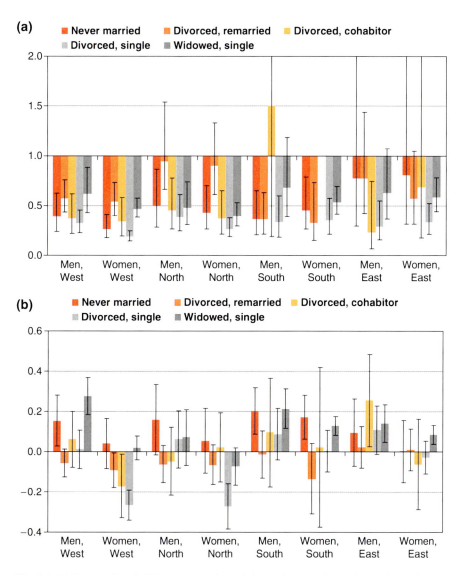

Fig. 9.4 (**a**) Impact of marital history categories, relative to those continuously married, on home-ownership, net of control variables, by gender and region (odds-ratios). (**b**) Impact of marital history categories, relative to those continuously married, on net wealth (log), net of control variables, by gender and region (regression coefficients)

widowhood seem to have a larger impact for women than for men. Coefficient point estimates indicate that these life events seem to have especially negative effects on net wealth for women in the South. For those countries, and also for those of the East, the small number of people in the samples who experienced divorce makes it difficult to identify these effects.

Above, we hypothesised that high female labour market participation would be a variable capturing welfare state influences that may mitigate the impact of marital dissolution for women. In order to test this hypothesis, we estimated separate regressions for each country (for women only), and plotted the coefficient estimates for the categories "divorced, single" and "divorced, remarried" against the female labour market participation rate in 1980. Figure 9.5 shows the results for net wealth (for home-ownership no relationship emerges in the plots). As regards the coefficient for the category "divorced, remarried" (Fig. 9.5b), we see no pattern. When

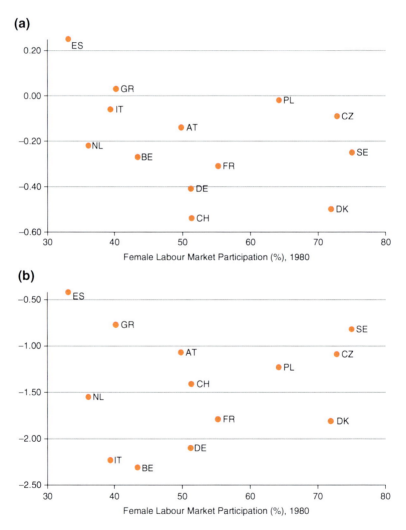

Fig. 9.5 (a) Impact of divorce on net wealth of being divorced and single, by female labour market participation (country level). (b) Impact of divorce on net wealth of being divorced and remarried, by female labour market participation (country level)

considering the impact on net wealth of being "divorced, single" (Fig. 9.5a), relative to being continuously married, it seems that in countries where labour market participation is high, e.g. Sweden, this effect is small. When labour market participation is lower, the impact can be much more substantial (e.g. Germany, Italy), but it can also be quite limited (e.g. Spain). The latter may be due to an "elitist" pattern of divorce, where only women who can afford to do so, having sufficient income of their own, take the step to a marital separation. In any case, the possible impact of this and other context variables needs to be analysed in a more rigorous way.

9.6 Conclusion

The goal of this study was to examine the impact of marital dissolution at some point during the the life course, particularly divorce, on the wealth of elderly people in Europe. A secondary aim was to assess whether the impact of different marital trajectories would differ across the different welfare state regimes. Our analyses indicate that persons who experienced a divorce have less wealth and lower chances of being home-owners than continuously married persons. Although remarried and cohabiting individuals who have ever experienced a divorce have higher levels of wealth and higher chances of owning the house they are living in, compared to single individuals, we find a lasting effect in old age of marital separation in many countries. This is an interesting result, which could only be obtained by looking at the life histories of individuals. Most research on this topic to date has only considered current marital status, and few studies have looked at the long-term consequences of marital dissolution.

In many countries, but not all, we find larger effects of marital dissolution for women than for men, as expected. We also found mixed support for our hypothesis that there would be large differences between European countries in the impact of marital separation on the wealth of elderly. However, the differences are not easily interpretable in terms of welfare state typologies, or specific welfare state interventions. While the hypothesised mitigating effect of high female labour market participation emerged to some extent from the results, more work is needed to identify such effects with a reasonable degree of certainty.

References

Andreß, H.-J., Borgloh, B., Bröckel, M., Giesselmann, M., & Hummelsheim, D. (2006). The Economic consequences of partnership dissolution. A comparative analysis of panel studies from Belgium, Germany, Great Britain, Italy and Sweden. *European Sociological Review, 22* (5), 533–560.

Dewilde, C. (2008). Divorce and the housing movements of owner-occupiers: A European comparison. *Housing Studies, 23*(6), 809–832.

Dewilde, C. (2009). Divorce and housing: A European comparison of the housing consequences of divorce for men and women. In H.-J. Andreß & D. Hummelsheim (Eds.), *When marriage ends. Economic and social consequences of partnership dissolution*. Cheltenham: Edward Elgar.

Dewilde, C., & Uunk, W. (2008). Remarriage as a way to overcome the financial consequences of divorce. A test of the economic need-hypothesis for European women. *European Sociological Review, 24*(3), 393–407.

Halvorst, R., & Palmquist, R. (1980). The interpretation of dummy variables in semilogarithmic equations. *American Economic Review, 70*(3), 474–475.

International Labour Organization (ILO). (2010). *Key indicators of the Labour Market Programme*. Retrieved June 5, 2010, from http://kilm.ilo.org/KILMnetBeta/default2.asp

Uunk, W. (2004). The economic consequences of divorce for women in the European Union: The impact of welfare state arrangements. *European Journal of Population, 20*, 251–285.

Wilmoth, J., & Koso, G. (2002). Does marital history matter? Marital status and wealth outcomes among preretirement adults. *Journal of Marriage and Family, 64*(1), 254–268.

Part II
Work and Retirement

Chapter 10
Early and Later Life Experiences of Unemployment Under Different Welfare Regimes

Martina Brandt and Karsten Hank

10.1 Unemployment in a Life Course and Cross-National Perspective

Involuntary job loss has been shown to be associated with a variety of adverse outcomes, such as lower wages, poorer health, or greater divorce risks (Arulampalam 2001; Hansen 2005; also see Chap. 17). It is thus considered as a serious life-disrupting event which may affect – in different ways – labour market entrants as well as older workers (Breen 2005; Henkens et al. 1996).

A much discussed issue of high policy relevance is the degree to which the experience of unemployment inflicts longer term "scars", such as an increased likelihood of future unemployment (Gangl 2004). Due to scarring effects, the total costs of unemployment might be higher than the immediate loss of earnings, thereby increasing lifetime inequality. To address this, one obviously needs to take a life course perspective – which is also important to account for the potential role of childhood conditions (such as parental socio-economic status) in predicting involuntary job loss in adulthood (Caspi et al. 1998).

Moreover, institutional factors – which are often specific to particular welfare state regimes – matter greatly for our understanding of unemployment at various stages of the life course (Blossfeld et al. 2006; Breen 2005). An issue of particular concern here is the extent to which welfare state interventions, such as active labour market policies, can mitigate longer term scars of previous unemployment experiences (Gangl 2004; Strandh and Nordlund 2008).

Exploiting retrospective information collected in the SHARELIFE project, we track the unemployment experiences of today's elders in 13 Continental European

M. Brandt (✉)
Mannheim Research Institute for the Economics of Aging (MEA), L13, 17, 68131 Mannheim, Germany
e-mail: brandt@mea.uni-mannheim.de

K. Hank
Institute of Sociology, University of Cologne, Greinstr. 2, 50939 Cologne, Germany
e-mail: hank@wiso.uni-koeln.de

countries from their labour market entry to retirement, addressing the following questions:

How did levels of unemployment in the current SHARELIFE sample vary across different stages of individuals' life course? What does the cross-national pattern of unemployment levels in early-, mid- and later-life look like?

How are childhood conditions associated with unemployment experiences across the adult life course? Are potential associations stronger in the beginning of individuals' employment career, fading-out later on?

To what extent do we observe scarring effects of unemployment – and variations therein across different welfare regimes – in Continental Europe's contemporaneous older population?

10.2 Unemployment Histories in SHARELIFE: Measurement and Analysis

SHARELIFE provides annual work histories of all respondents who reported to have ever done any paid work from the year when they left full-time education till the date of the interview (also see Chap. 11). The data allows us to identify periods of unemployment only, if the gap between two jobs (between leaving full-time education and the respondent's first job, respectively) was longer than 6 months. We restrict our analysis to respondents aged 60 years or older at the time of the SHARELIFE interview. Cases for which the interviewer reported frequent difficulties to understand the questions being asked are excluded from the analysis. This results in a sample of roughly 16,000 observations.

Our main variables of interest are three binary indicators of unemployment periods (any vs. none) during different phases of the individual's career: early (i.e. within the first 3 years after having left full-time education), prime-age working years (between the end of "early" and age 49), late (age 50 or over). We use these variables to perform separate analyses of unemployment risks in each career phase. Moreover, information on individuals' previous experience of involuntary job loss enters the models for unemployment during prime-age and later years as an explanatory variable. We interpret significant associations of unemployment in t-1 with the risk of experiencing a period of unemployment in phase t of one's career as indication for scarring effects.

All control variables refer to measures which were determined before the individual first entered the labour market. That is, we exclude potentially relevant factors which themselves might have been affected by earlier experiences of unemployment, such as mid-life health (Chap. 17) or marital status (Hansen 2005). Next to sex and years of education (derived from SHARE's Wave 1 & 2) as basic socio-demographic variables, we focus on information regarding respondents' childhood conditions at age 10: self-rated health (ranging from poor [1] to excellent [5]), a composite measure of self-rated maths and language skills (ranging from "much better" [1] to "much worse" [5] in comparison to other children in the respondent's class), and parents' socio-economic status (operationalised by the number of rooms per person and the number of books available in the household).

Finally, we roughly account for the historical period during which the SHARE-LIFE respondents graduated (pre-1950, 1950s, 1960 or later) as well as for the respective welfare state context. Starting from Esping-Andersen's (1990) initial "regime" typology we roughly group the countries represented in SHARELIFE into four clusters which we label as social-democratic (DK, NL, SE), conservative (AT, BE, CH, DE-W, FR), Mediterranean (ES, GR, IT), and post-communist (CZ, DE-E, PL). Moreover, in a supplementary analysis we take into consideration the labour market situation in each country (except the post-communist ones) by accounting for the level of unemployment in the year when respondents' left education (Kahn 2010).

10.3 Regional and Career Stage Variations in Levels of Unemployment

While differences in unemployment levels at various stages of individuals' life-course (career, respectively) are rather modest on average (i.e. if the pooled SHARE-LIFE sample is considered), we observe substantial variations across different welfare state regimes (see Fig. 10.1). Consistent with previous research we find the by far highest prevalence of early career unemployment during the post-WWII era in the Mediterranean countries (slightly more than 12%), which, however,

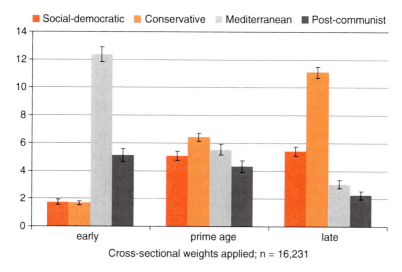

Fig. 10.1 Prevalence of unemployment (in %) at different life-course/career stages across welfare state regimes

exhibit the lowest unemployment rate – together with the post-communist countries – among older workers (somewhat less than 3%). The reverse is true under the conservative and social-democratic regimes, where unemployment in the first years after leaving education is a relatively minor issue (~2%), but affects more than 8% of workers aged 50+ in the conservative and 6% in the social-democratic countries.

10.4 Correlates of Job Loss Over the Life Course

We estimated logistic models for three "events", namely the experience of at least one unemployment spell during the individual's early, mid-, and late career (see Table 10.1), whose results we discuss jointly.

In the early unemployment model (Model 1), the number of rooms per member of the parental household was the only personal characteristic bearing a statistically significant correlation with the dependent variable. More rooms (supposed to reflect a higher parental SES) were associated with substantially lower odds of being unemployed in the first years after the individual's exit from full time education. This relationship holds for unemployment risks later in life, too. Our composite measure of math and reading skills as well as the number of books in the parental household exhibit the same – but insignificant – outcome in all models. The coefficient of our last measure of childhood conditions, subjective health, was insignificant in the "early unemployment" model, but suggests that better health at age 10 is associated with significantly lower odds of unemployment later in life.

The outcome of other correlates of job loss tends to be quite different before and after age 50, though, both in terms of their sign and statistical significance. Women, for example, are more likely than men to report unemployment before age 50 (probably due to difficulties finding a family friendly job), but they are less likely to be unemployed after age 50 (probably due to an earlier selection of less career oriented women out of the labour force). More years of education significantly reduce the odds of being unemployed before age 50, but the coefficient is not statistically significant anymore if job losses after age 50 are considered (suggesting that human capital obtained earlier in life might loose at least some of its "protective" effect over time).

Turning to our cohort and welfare regime indicators, we – again – observe very similar associations in the models for mid- and late-career unemployment. Those who graduated after 1950, and particularly those who completed their education in the 1960s or later, exhibit significantly higher odds of losing their job than those who left education before 1950. Individuals living under a Mediterranean or post-communist welfare regime face much smaller unemployment risks in their mid- and late-career, compared to their counterparts in conservative countries. The same holds for people from social-democratic countries, but note that the respective coefficient in the "mid-career unemployment" model does not meet the 10% – level of significance. A very different picture, however, emerges when looking at the "early unemployment" model. There are no cohort differences, individuals from

Table 10.1 Odds ratios from logistic regressions for "early", "prime-age", and "late" unemployment

	Model 1: "Early"	Model 2: "Prime-age"	Model 3: "Late"
Gender (female)	0.89	1.50**	0.67**
	(1.38)	(5.40)	(5.56)
Years of education	1.01	0.95**	1.00
	(0.83)	(4.72)	(0.20)
Childhood conditions			
Number of rooms per person	0.59**	0.74*	0.78*
	(3.79)	(2.24)	(2.49)
Maths and reading skills	0.92	0.95	0.99
	(1.42)	(0.91)	(0.15)
Number of books in household	0.96	0.96	0.95
	(0.93)	(0.98)	(1.57)
Self-rated health (good or better)	1.00	0.91*	0.92*
	(0.11)	(2.43)	(2.42)
Cohort: left education ...			
... Before 1950	1.00	1.00	1.00
... 1950–1959	0.91	1.70**	1.70**
	(1.00)	(4.97)	(5.13)
... 1960 or later	0.95	2.83**	2.15**
	(0.42)	(9.59)	(7.15)
"Welfare regime"			
Social-democratic	0.66**	0.87	0.68**
	(2.74)	(1.60)	(4.59)
Conservative	1.00	1.00	1.00
Mediterranean	4.69**	0.56**	0.27**
	(14.34)	(5.31)	(10.79)
Post-communist	1.14	0.40**	0.21**
	(0.89)	(6.42)	(9.96)
Previous unemployment			
"Early"	–	2.83**	1.42*
		(7.57)	(2.06)
"Prime-age"	–	–	3.00**
			(10.15)
Pseudo-R2	0.09	0.04	0.06

z-values in parentheses; robust standard errors
n = 15,913
*Significant at 5%; **Significant at 1%

conservative and post-communist regimes do not differ from each other, whereas the odds of job loss are particularly high in the Mediterranean and lowest in the social-democratic countries.

Finally, we find clear indication of scarring effects. There is a large positive total effect of having experienced unemployment in the first 3 years of one's career on further unemployment risks before and after age 50. Although the odds ratio of our "early unemployment" indicator in the late-career model (Model 3) is only half the size of the respective odds ratio in the mid-career model (Model 2), the observed correlation even remains highly significant, if unemployment during the prime-age working years (which is also significant) is controlled for in the regression. That is, early- and mid-life experiences of unemployment bear in them independent and strong associations with the risk of losing one's job after the age of 50.

Additional models employing interactions of "early" unemployment with gender, cohort, and welfare regimes (details not shown) generally did not provide any additional insights. However, scarring effects among prime-age respondents living in "social-democratic" and "post-communist" countries were found to be significantly larger – by a factor of 2 to 3 – than among their counterparts elsewhere.

10.5 Unemployment as a Contextual Variable: Does It Matter?

One possible explanation for this latter finding has been put forward by Lupi and Ordine (2002), who argue that in a high youth unemployment environment such as Southern Europe individual unemployment experiences tend to be perceived as "normal" and do not necessarily signal poor quality of the worker, as is the case in Northern Europe, where youth unemployment is not an issue of general concern.

To test, whether there is support for this argument in the SHARELIFE sample, we estimated a model in which we controlled for country-level unemployment rates at the time of individuals' exit from education (excluding the formerly communist countries). Since, for the entirety of countries contributing to our sample, this contextual variable is only available from 1960 onwards, we were forced to restrict this part of the analysis to those who left school in that year or later. While we find the expected positive association between individuals' risk of "early" unemployment and the national unemployment rate in the year when they exited from education, the respective interaction term turned out to bear no statistically significant association with later-life experiences of unemployment (see Table 10.2).

10.6 Summary and Perspectives for Future Research

SHARELIFE allowed us to observe interactions between differences in unemployment across individuals' life-course and across welfare state regimes where, for example, early (late, respectively) unemployment is relatively high (low, respectively) in the Mediterranean countries, whereas the reverse pattern holds under a conservative welfare regime. While the results of our multivariate analysis suggest that welfare state institutions matter at all stages of the individual's employment career, they appear to be particularly relevant in the first 3 years following the exit from full time education. Different from later career phases, the personal characteristics considered in the "early unemployment" model barely matter, but we find strong positive (Mediterranean) and negative (social-democratic) associations of our welfare regime indicators with younger individuals' unemployment risks. Both labour market institutions as well as elements of a country's educational system, such as employment protection regulations or the vocational specificity of the

Table 10.2 Odds ratios from logistic regressions for "early", "prime-age", and "late" unemployment in cohorts who left education in 1960 or later

	Model 4: "Early"	Model 5: "Prime-age"	Model 6: "Late"
Previous unemployment			
"Early"	–	4.04**	0.82
		(2.89)	(0.34)
"Prime-age"	–	–	3.4**
			(6.83)
Context variable			
Unemployment rate in year of exit from education (1960+)	1.6**	0.94	0.77**
	(13.24)	(1.47)	(4.48)
Micro–macro interaction			
Unemployment rate X Early unemployment experience	–	0.87	1.17
		(1.04)	(1.03)
Pseudo-R2	0.12	0.04	0.05

Notes: Post-communist countries and respondents who left education before 1960 are excluded. All variables displayed in Table 10.1, except welfare regime, are control-led for in the model. Unemployment rates (in %), by country and year, retrieved from: http://www.fgn.unisg.ch/eumacro/macrodata/macroeconomic-time-series.html [05.04.2010]
z-values in parentheses; standard errors clustered by country–year
$n = 4,851$
*Significant at 5%; **Significant at 1%

education system (Breen 2005), are likely to play an important role here and deserve further attention in future research.

For mid- and late-career workers, the odds of losing one's job bear significant correlations with gender, years of education, and cohort. To some degree, childhood conditions (such as parental SES and the respondent's health) also seem to matter, but the underlying causal mechanisms are not yet well-understood and clearly require further investigation.

Our analysis of SHARELIFE data confirms previous research in that we find clear evidence for scarring effects, even among older workers. We can show that early- and mid-life experiences of unemployment bear in them independent and strong associations with the risk of losing one's job after the age of 50. This observation holds for men as well as for women and is stable across school-leaving cohorts and welfare regimes.

This initial analysis of SHARELIFE is far from being comprehensive and the perspectives for future research are manifold, e.g.:

- How much of the "total effect" of previous unemployment experiences identified here is mediated through mechanisms related to adverse effects of involuntary job loss on individuals' health (Chap. 17) or marital status (Hansen 2005), for example?
- What evidence does SHARELIFE provide for scarring effects of unemployment on life-time earnings (Arulampalam 2001)?

- Further efforts are needed to understand more thoroughly the role of specific welfare state interventions (Gangl 2004) and the impact of macro-economic conditions (Lupi and Ordine 2002) in the observed variations of unemployment – and its scarring effects – across cohorts and countries. Why, for example, are scarring effects among Scandinavian prime-age workers stronger than in the Mediterranean countries?

References

Arulampalam, W. (2001). Is unemployment really scarring? Effects of unemployment experiences on wages. *The Economic Journal, 111*, F585–F606.

Blossfeld, H.-P., Buchholz, S., & Hofäcker, D. (Eds.). (2006). *Globalization, uncertainty and late careers in society*. London: Routledge.

Breen, R. (2005). Explaining cross-national variation in youth unemployment. Market and institutional factors. *European Sociological Review, 21*, 125–134.

Caspi, A., Entner Wright, B. R., Moffit, T. E., & Silva, P. A. (1998). Early failure in the labor market: Childhood and adolescent predictors of unemployment in the transition to adulthood. *American Sociological Review, 63*, 424–451.

Esping-Andersen, G. (1990). *The three worlds of welfare capitalism*. Princeton: Princeton University Press.

Gangl, M. (2004). Welfare states and the scar effects of unemployment: A comparative analysis of the United States and West Germany. *American Journal of Sociology, 109*, 1319–1364.

Hansen, H. T. (2005). Unemployment and marital disruption – A panel data study of Norway. *European Sociological Review, 21*, 135–148.

Henkens, K., Sprengers, M., & Tazelaar, F. (1996). Unemployment and the older worker in the Netherlands: Re-entry into the labour force or resignation. *Ageing and Society, 16*, 561–578.

Kahn, L. B. (2010). The long-term labor market consequences of graduating from college in a bad economy. *Labour Economics, 17*, 303–316.

Lupi, C., & Ordine, P. (2002). Unemployment scarring in high unemployment regions. *Economics Bulletin, 10*, 1–8.

Strandh, M., & Nordlund, M. (2008). Active labour market policy and unemployment scarring: A ten-year Swedish panel study. *Journal of Social Policy, 37*, 357–382.

Chapter 11
Labour Mobility and Retirement

Agar Brugiavini, Mario Padula, Giacomo Pasini, and Franco Peracchi

11.1 Work Life Histories

Public and private pensions represent the main income source of Europeans in their old age. Pension provisions depend on workers' career: workers retiring under the same pension regime might receive very different pension benefits depending on the time pattern of employment and unemployment spells, on the magnitude and the timing of income shocks and, more generally, on the age-profile of earnings (Boeri and Brugiavini 2009). Individuals are often induced to retire early, but inadequate contributions due to a short work history may reduce household's welfare later in life (Angelini et al. 2009). Previous work has shown great variability in the patterns of transition from work to retirement (Peracchi and Welch 1994; Brugiavini and Peracchi 2005; Brugiavini et al. 2008), yet little is known about how the working history of individuals affects their resources in retirement. This would require panel data where workers are followed over a long period of time. Long panels, however, are hardly available and would be subject to serious problems of attrition and non-response. A simple alternative are cross-sectional data or short panels with a retrospective component, such as SHARE. The SHARELIFE component of this survey provides information on the history of past employment and allows accounting for periods of unemployment during the working life. In this paper we relate incidence and duration of unemployment, as well as tenure in a job, to the replacement rate at retirement. The international comparability of the data provides insights on the effect of labour market institutions and regulation on workers' mobility, and on the insurance properties of social security against earnings risks.

A. Brugiavini (✉), M. Padula, and G. Pasini
Dipartimento di Scienze Economiche, Università Ca' Foscari Venice, Cannaregio, 873, 30121 Venice, Italy
e-mail: brugiavi@unive.it; mpadula@unive.it; giacomo.pasini@unive.it

F. Peracchi
Facoltà di Economia, Università Tor Vergata, via Columbia, 2, 00133 Roma, Italy
e-mail: franco.peracchi@uniroma2.it

The rest of the paper is organized as follows. Section 11.2 documents the patterns of jobs turnover across European countries. The data show marked differences between Northern and Southern Europe: in Denmark, at one end of the spectrum, the average worker changes six jobs in his working life, in Greece, at the other end of the spectrum, only once. The same number of transitions between jobs during a working life may result in very different contribution histories, depending on the tenure at each job and the duration of unemployment spells between jobs. Therefore, Sect. 11.3 reviews duration data and, again, provides evidence of a North–South gradient. Job tenure and mobility feed back into replacement rates, which are investigated in Sect. 11.4. The data reveal smaller cross-country differences in replacement rates, suggesting that the design of social security systems may in fact contribute to attenuate the effect of different career patterns on retirement benefits. Section 11.5 concludes.

11.2 Job Turnover

In this section we document the wide variability in working histories of European workers. We start by looking at the number of jobs that individuals held in their life. Figure 11.1a, b is based on the entire sample and show that women tend to have a lower number of jobs, particularly in Poland and in Southern countries such as Italy, Spain, and Greece. The German sample is split into those who in 1989 were living in East Germany (DE-E) and those who were in West Germany (DE-W). Irrespective of gender, younger cohorts have a higher number of jobs, indicating that mobility between jobs is more common for the younger old. Note that some of the younger old are still working, so that our data may effectively underestimate the degree of labour mobility The gender gap is smaller in Northern than in Southern countries, where female participation is lower. Differences across cohorts are also smaller in Northern countries. The overall picture is one of greater job turnover in Northern countries, where the differences between genders and across cohorts are generally smaller. Depending on the social security system and the pension legislation, higher mobility may be associated with "losses" in terms of pension and social security benefits, for example if pension rights are not fully portable or if changes are associated with long unemployment spells. It should be stressed that mobility per se does not necessarily have a negative effect on pensions, as for example workers may go from a low-pay job to a high-pay job hence increasing their pension levels. It is the existence of distortions in the pension rules (such as lack of portability) which might create these losses in association with mobility (Gustman and Steinmeier 1993).

Moreover, even when pension benefits are fully portable, pension income may vary depending on whether job turnover is associated to job-to-job transition or to transition from job to long unemployment spells. Therefore, in Fig. 11.2a, b we graph the average number of job changes which resulted in at least 6 months out of employment, again grouping the data by gender and year of birth. The data show

11 Labour Mobility and Retirement 127

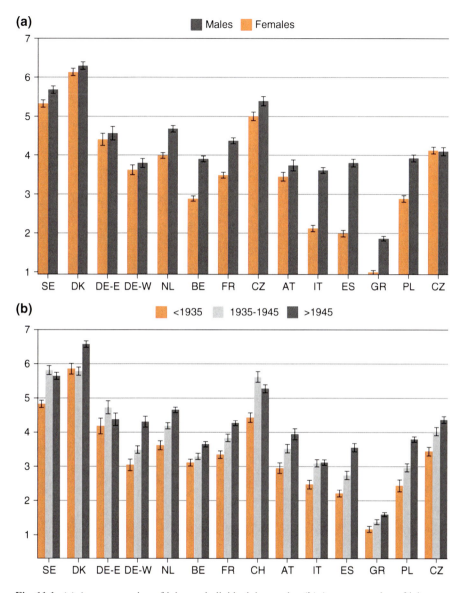

Fig. 11.1 (**a**) Average number of jobs per individual, by gender. (**b**) Average number of jobs per individual, by cohort

differences between countries with Northern countries and East Germany now scoring a higher number of job-to-unemployment transitions compared to Southern European countries. This is particularly true among women, and as a consequence gender differences are wider in Scandinavian rather than Southern countries. Figure 11.2a, b also indicate that most of the job changes during the life span are job-to-job transitions, which are unlikely to imply losses in terms of pension

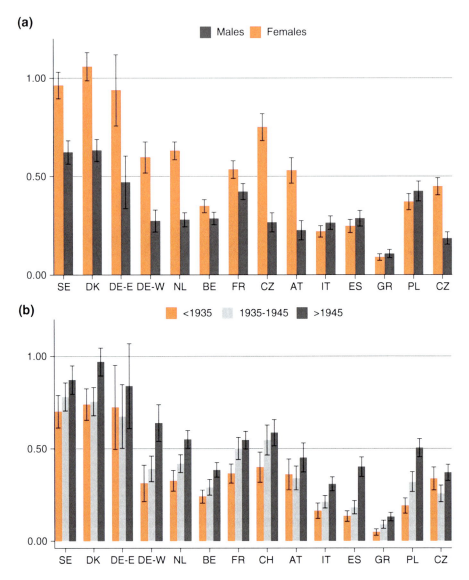

Fig. 11.2 (a) Average number of job changes per individual which entailed 6 months or more out of employment, by gender. (b) Average number of job changes per individual which entailed 6 months or more out of employment, by cohort

benefits: the average number of changes which involved at least 6 months out of employment is much lower than the average number of jobs per individual.

In terms of gender differences, European countries can be split into two groups. In the first group (Sweden, Denmark, East and West Germany, the Netherlands, Switzerland, Austria and the Czech Republic) women have approximately twice the number of job changes than men, which suggests that women experience many transitions out-of-the-labour-market. In the remaining countries (Italy, Spain, Greece, France, Belgium and Poland) gender differences are smaller. This does not necessarily indicate that labour market conditions for women are better: it is quite likely that in these countries women experience fewer employment episodes, and thus fewer transitions from one job to another. Differences across cohorts are almost not significant in Sweden, while for Spain, at the other end of the spectrum, the average number of job changes which involved more than 6 months out of employment is twice as large for the youngest old, as shown in Fig. 11.2b. Furthermore, differences across cohorts are larger in Fig. 11.2b than in Fig. 11.1b, probably due to changes in the employment protection legislation across cohorts depending on the time of entry to the job market.

Working histories might be similar in terms of job turnover, but very different when it comes to their effects on replacement rates. This depends on the portability of benefits between jobs, but also on average job tenure and the duration of unemployment spells between jobs. We therefore devote the next section to document how job tenure and unemployment duration vary across European countries.

11.3 Employment and Unemployment Duration

The SHARELIFE survey allows us to measure the length of time spent in each specific labour market state, i.e. the duration of the different spells of employment and unemployment. Being more specific, the respondent is asked to report each employment and unemployment episode lasting 6 months or more. Short employment and unemployment episodes are therefore excluded from our analysis.

Table 11.1 describes the distribution of duration of employment spells lasting at least 6 months. Therefore, the sample includes all the employments spells of any

Table 11.1 Percentage distribution of employment duration by country

	Duration in years			Number of observations
	0–24	25–34	35+	
SE	90.78%	5.07%	4.15%	6,898
DK	93.90%	3.89%	2.21%	8,669
DE-E	89.71%	5.73%	4.56%	1,030
DE-W	84.82%	7.86%	7.32%	2,213
NL	91.34%	5.15%	3.51%	6,353
BE	83.18%	9.56%	7.26%	5,994
FR	87.02%	7.96%	5.03%	6,146
CH	92.74%	4.03%	3.24%	4,420
AT	81.70%	10.40%	7.90%	2,481
IT	80.52%	11.51%	7.97%	4,441
ES	80.16%	8.48%	11.36%	3,608
GR	55.47%	18.59%	25.94%	2,340
PL	81.21%	11.31%	7.48%	4,093
CZ	77.39%	9.62%	12.98%	4,052
Total				62,738

individual who reported to have worked at least once in his/her life. The distribution is left skewed for all countries, with marked differences across countries: Denmark displays the highest degree of left-skewness, while for Greece the distribution is relatively flat. Jobs at different points in a working life are likely to have different tenures, due to search for a good match at the beginning of the career and job-specific skill acquisition later in life. Since we can observe duration and timing of employment spells over the life cycle, we can study the distribution of employment spell changes between jobs, and in particular between the first job and the last job. The results, which we do not show for brevity, reveal interesting patterns: the distribution of employment duration for the last job is quite even across all European countries. This is perhaps not surprising, if the cost of searching for a good match increases with age at a similar rate across European countries.

It is worth noticing that the fraction of people reporting extremely long employment spells is somewhat higher than expected: 6.64% of the overall sample of spells lasts 35 years or more, 3.25% more than 40 years, and 1.21% lasts more than 45 years. This does not seem to be driven by employment duration of civil servants, which is likely to be on average longer than employment duration of private sector employees: the same table obtained excluding civil servants exhibit the same long right tail. Even after dropping spells longer than 45 years, as they are likely to be outliers, the distribution might give more weight to longer spells than expected, for three reasons: first, since we are dealing with older workers it is common to start working at very young ages and hence it is likely that these generations had long spells. Second as already explained spells shorter than 6 months are not considered, thus reducing the probability mass on the left tail. Third, interviewers were asked to instruct respondents to count short-term jobs which are similar, even if working for different employers, as a single spell and not record changes of job for the same employer as different spells.

Figure 11.3 shows the average duration of long unemployment spells in years. In all countries (except Czech Republic), women experience longer unemployment spells. Looking at the average duration of unemployment spells can however hide important differences between workers, beyond those simply captured by gender.

11.4 Replacement Rates

Depending on the pension regime a person faces at different stages of her life, the sequence of jobs in her work history could be more relevant or less relevant for her pension benefits entitlements and therefore for her replacement rate. In this section, we compute the replacement rate as the ratio of the first benefits received over the last wage in the main job for those respondents who indicate as their main job the one that ended with a transition into retirement. Note that in this way we neglect those who had a secondary job in the last years before retirement, those who chose other pathways to retirement (for example through disability), and those who experienced a long period out of employment before retirement, e.g. women who

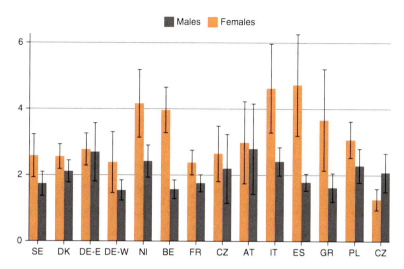

Fig. 11.3 Average number of years spent in unemployment conditional on having been unemployed

leave their job to take care of home and family. We also drop the self-employed and focus only on private and public sector employees: rules concerning benefit calculation are quite different for the employees and the self-employed, and for the latter it is often problematic to relate replacement rates to work histories.

Figure 11.4a shows the average replacement rate (the average of individual replacement rates). For most countries it ranges between 0.70 and 0.80, with Austria, Belgium and Spain featuring a lower rate, and East Germany and Greece a slightly higher rate. In Figs. 11.4b, c the observations are grouped by gender and year of birth. Changes in institutional arrangements which affect the generosity of the pension system may in principle lead to gender and cohort differences within countries. We find that these differences are not statistically significant.

Given the large heterogeneity of replacement rates across countries, it is important to understand to what extent the observed differences correlate with differences in number and duration of employment and unemployment spells.

Figure 11.5a, b show the relation between the replacement rate and the number and duration of employment spells respectively. The straight line is the regression line fitted to the data. Because East Germany and Greece appear as outliers, we include them in the scatter plot but not in the fitted regression lines. Figure 11.5a reveals a positive relation between the number of employment spells and the replacement rate. This may be due to the fact that, as we have already shown, most of the job changes are job-to-job transitions. Therefore, a higher number of employment spells may signal a faster career and a higher contribution accrual, rather than a difficulty in retaining a job. By the same token, a longer tenure does not have any statistically significant effect on replacement rates (see Fig. 11.5b): as long as most of the job changes are job-to-job movements, there is no reason to expect the average duration of an employment spell to matter in terms of pension rights.

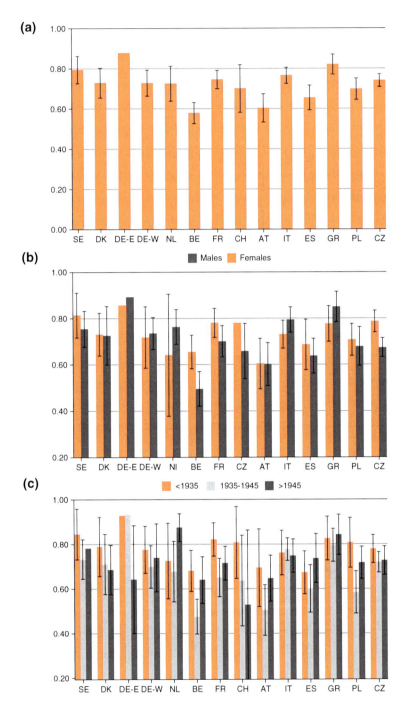

Fig. 11.4 (a) Replacement rate by country. (b) Replacement rate by gender. (c) Replacement rate by cohort

11 Labour Mobility and Retirement

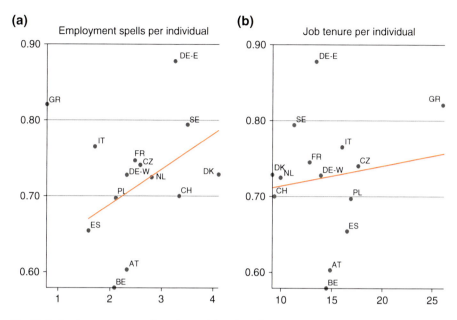

Fig. 11.5 Replacement rate and number and duration of long employment spells

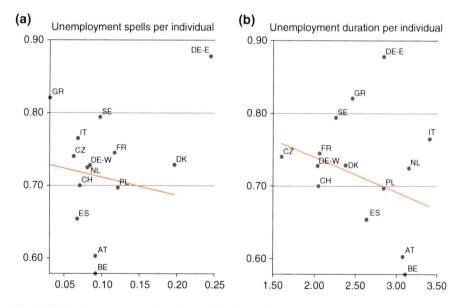

Fig. 11.6 Replacement rate and number and duration of long unemployment spells

Figures 11.6a, b show the scatter plots of the average number and the average duration of long unemployment spells by country against the average replacement rate. Although we include the observation for East Germany in the scatter plot, we exclude this country from the regression line, because its high average number of

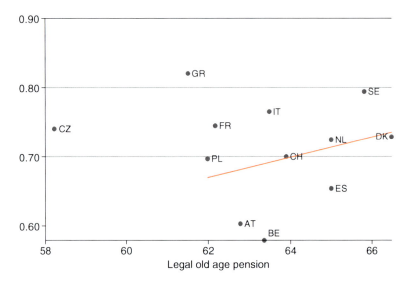

Fig. 11.7 Replacement rate and legal old age pension at the moment of retirement

unemployment spells may just be due to the very small sample size. The figures reveal a clear negative relationship between the replacement rate and the number as well as the duration of long unemployment spells. Countries characterized by work history with many and long unemployment spells tend to exhibit lower replacement rates.

Finally, Fig. 11.7 relates the actual average replacement rate of each country with the average legal retirement age prevailing at the time of retirement in the sample. Once again we exclude Greece and the Czech Republic from the regression line because the institutional data about the past legal retirement age and also about the actual replacement rate show some inconsistencies. The legal retirement age for old age pension is the country-average of the legal retirement age for each individual at the time of her retirement, which is meant to emphasize the role of institutions through the "exogenous" legal retirement age. The positive association between the actual replacement rate and the legal retirement age suggests that in those countries where people are induced to work longer (higher retirement age) they also achieve higher replacement rates. This result might be due to both a "formula" effect, i.e. it might depend directly on the way the length of the career enters the benefit calculation and a "filling" effect, i.e. individuals who have the opportunity to work longer may fill the contributions gaps due to unemployment spells.

11.5 Conclusions

The patterns of job mobility vary greatly across European countries. The number of jobs held during a career by the average worker is much higher in Northern than in Southern countries. The available evidence points into the same direction: labour markets in Southern countries are much less mobile than in Northern countries. However, cross-country differences in job mobility do not translate into corresponding cross-country differences in replacement rates. On the other hand, pension entitlements appear to be very sensitive to the number and duration of unemployment spells in the workers' careers.

Our results suggest that labour mobility, if anything, positively affects the replacement rate, and that institutions which help keep unemployment spells short, for example through active labour market policies, also reduce differences between less and more mobile workers. Therefore, the current tendency to deregulate European labour markets is not necessarily detrimental for income after retirement, provided it is tempered by policies designed to limit long-term unemployment.

References

Angelini, V., Brugiavini, A., & Weber, G. (2009). Ageing and unused capacity in Europe: Is there an early retirement trap? *Economic Policy, 59*, 463–508.

Boeri, T., & Brugiavini, A. (2009). Pension reforms and women retirement plans. *Journal of Population Ageing, 1*, 7–30.

Brugiavini, A., Pasini, G., & Peracchi, F. (2008). Exits from the labour force. In A. Börsch-Supan, A. Brugiavini, H. Jürges, A. Kapteyn, J. Mackenbach, J. Siegrist, & G. Weber (Eds.), *Health, ageing and retirement in Europe (2004–2007), starting the longitudinal dimension* (pp. 204–212). Mannheim: MEA.

Brugiavini, A., & Peracchi, F. (2005). The length of working lives in Europe. *Journal of the European Economic Association, 3*, 477–486.

Gustman, A., & Steinmeier, T. (1993). Pension portability and labor mobility: Evidence from the survey of income and program participation. *Journal of Public Economics, 50*, 299–323.

Peracchi, F., & Welch, F. (1994). Trends in labor force transitions of older men and women. *Journal of Labor Economics, 12*, 210–242.

Chapter 12
Atypical Work Patterns of Women in Europe: What Can We Learn From SHARELIFE?

Antigone Lyberaki, Platon Tinios, and George Papadoudis

12.1 Introduction

The second half of the twentieth century was a time of rapid social transformation. Nowhere were the changes more radical than in women's participation in society and work. Women increasingly claimed a fuller and more active position in all societal functions. Though all parts of Europe and all social strata were affected, this process was unevenly distributed over time and space and driven by a variety of influences. Such influences could have been structural changes in production, transformations in the function of the family and last, but not least, attitudes in what woman's position ought to be, as reflected in shifts of policy priorities. This period of rapid change corresponds to the lifetime of individuals in the SHARE survey. When today's 50+ population were young girls, the world they were entering was very different from today. The long term social changes correspond to lived experience of women in the SHARE sample. The women in SHARE were witnesses to the foundation, flowering and restructuring of the Welfare State. Social policy stances towards maternity and family policy as well as labour market institutions were defining fissures between certain forms of the so-called "European Social Model". This paper begins exploring how these factors – labour and social policy transformation – were reflected in the lives of women in the SHARELIFE sample.

For purposes of clarity of exposition, this paper utilizes the device of examining groups whose characteristics place them in a minority in their own country, yet which are very similar to majorities in other countries in the SHARE sample. Thus family-centred women who have never worked are the exception in Scandinavia,

---

A. Lyberaki (✉) and G. Papadoudis
Department of Regional and Economic Development, Panteion University of Social and Political Science, 136, Syngrou Avenue, Athens 17671, Greece
e-mail: antiglib@gmail.com; gpapadoudis@gmail.com

P. Tinios
Piraeus University, Karaoli & Dimitriou Street 80, Piraeus 18534, Greece
e-mail: ptinios@otenet.gr

yet are strongly represented in the South. Conversely career women in the South are uncommon, yet are the majority in the North.

These types of comparisons are useful for fixing ideas and for representational purposes. They can also be held to pose complex questions with clarity (approximating in logic a controlled experiment): Given that the kind of obstacles to employment which are held responsible for low labour participation in the South (child care facilities, income support) were patently available in the North, yet the minority chose traditional roles, what were the factors still placing obstacles to their participation? Was it limited availability of service infrastructure, a question of values, ill health or due to the vestiges of sex discrimination and insufficiency of financial incentives? Conversely, given that the shortcomings of social services are deemed sufficient to explain persistence of traditional roles for the majority in the South, how did career women cope with the pressures of balancing work and family? Did they have fewer children, did they have access to child care from family resources, or were they forced to work by financial pressure? How did women's own (socially conditioned) preferences affect their choices? Once we try to control for other factors, did social policy lead or follow developments?

12.2 Identifying the Groups: Dominant and Atypical Patterns

Patterns of female paid work vary hugely in Europe, as do work-care models. Evolving "models of family" (i.e. the shift away from the male breadwinner model in the direction of dual-earner families – Jane Lewis, 2001) and "preferences" (home-, work-centred or adaptives – K. Hakim 2000, 2004) have been ways of analyzing complex trends. At the same time, economists have noticed the existence of two ideal-types which can be rationalized as the result of two equilibria in Europe regarding women's work patterns: a high labour force participation, good social infrastructure and high fertility rates equilibrium characterizing Northern countries, and a low participation, low fertility and missing social infrastructure equilibrium characterizing Southern economies (Boeri 2003; Boeri et al. 2005; Bettio and Villa 1998). This brings the welfare state into the discussion as an important influence. Esping-Andersen's welfare state typology leads one to expect patterns of female paid work to observe the boundaries of the "Worlds of Welfare Capitalism". The exact typology of Esping-Andersen has been questioned in the context of gender – Lewis et al. (2008). However, this criticism does not affect the geographical dividing lines which largely remain. The effect of the welfare state can be brought to bear in the explanation of inter-country and individual differences, or might point at specific areas of intervention. As Daly (2002) states in reviewing the current state of knowledge, evidence on a direct link between policies and particular female labour profiles is still inconclusive. SHARELIFE, by providing data on the entire life of respondents (rather than synchronic information)

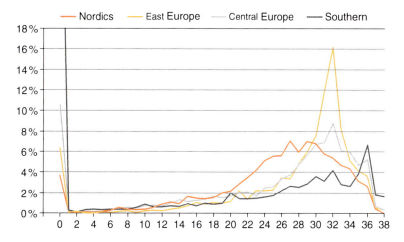

Fig. 12.1 Distribution of career length to age 50, by country group

allows us to introduce a time and cohort dimension. We can thus see not only whether patterns exist, but how they spread through time.

In defining female work patterns a number of ideal-types stand out; these are usually associated with the country groups in which they are prevalent. Hakim's (2000) work predisposes to find women distributed in clusters around these behaviour norms. Our first concern is to identify whether such groups exist. In doing so, the simple expedient of comparing years worked introduces bias, as older respondents will include years of work after 50 and will systematically exhibit longer careers than 50-year olds who are still working. To allow for this, the key variable to be analysed is years of work of each respondent until the age of 50 – regardless of age. Figure 12.1 shows the distributions of this variable for the four geographical groups which roughly correspond to distinct types of welfare state: The North, Centre, South and East.

Simple visual inspection shows the existence of two polar types. The "Full career woman" (FCW) or work-centred woman. In Fig. 12.1 we see concentrations of women with around 30 years of work or more, which with an entry age of 20 essentially implies uninterrupted stay in employment (for those women with tertiary education, a full career necessarily starts later, so the FCW category is defined to include those with more than 26 years work) . At the other extreme, a large group of women never worked at all – the "full-time carer" or "family-centred" woman – full family woman (FFW) with no links to the labour market. Hakim's category of the "adaptive" woman falls in between (supplemental earner, main carer, in and out of work). This category can be further divided according to work-intensity (i.e. share of working years in total). For the purposes of exposition the continuum is divided into two groups: between 20 and 29 years "Adaptive Career Woman" (ACW) and, between 1 and 19 years of work "Adaptive Family Woman" (AFW). What distinguishes the one from the other is the different degree of continuity of

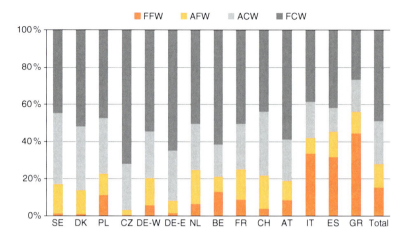

Fig. 12.2 Women's career pattern by country

employment characterizing the two groups. The picture of Fig. 12.2 largely confirms Hakim's expectation and leads to the following classification by country.

Figure 12.2 examines the groups by country. Never worked group: The Southern countries are the champions (over 30%, on average, with Greece reaching almost 45%), while only a small minority (below 6%) in five countries (Denmark and Sweden (the Nordics), Czech Republic, Germany and Switzerland), followed by France, Austria and Poland (around 10%). By contrast, the longest careers are recorded in the Nordics (Sweden and Denmark), the Czech Republic (and also former GDR), with over 60% of women working longer than 31 years. Long careers are also the rule in Central countries (ranging from 40 to 50%) – in Austria, Germany, France, Poland and Switzerland and Belgium at the limit. The group of adaptives is also largest in the North, signifying the possibility of leaving and then being able to return to work. The Mediterranean countries and the Netherlands, on the other hand, have moderate share of working women in long careers, and few adaptives.

How does this picture change by cohort? If we examine the career length by country group we can observe that in all cases there is an actual decrease of the never-worked FFW group. As time passes we generally see a decrease in polarization in favour of the adaptive group – with more than 19 years employment (ACW). We also see a reduction of very long careers (35+), connected presumably with later entry into the labour market as a result of the raising of the school-leaving age.

Examining the distributions, we may retain three crucial observations:

- The critical decision – taken early on – is whether to enter the labour market.
- Those entering the labour market appear in many countries most likely to continue on for a full career.
- The intermediate group – those exiting and re-entering the labour market – are in a minority, though their prevalence is higher in younger cohorts.

Hence it is important to take a closer look at career interruption patterns.

12.3 Career Interruptions of Women with Some Work Experience and Children

The crucial factor in women's working lives is childbearing. Here we focus only on mothers who had been working when they gave birth, i.e. x% of women with some work experience. We first show whether the arrival of a child affects working patterns (Table 12.1). The rule is to stop work temporarily (more than half of working women), although the share of permanent drop outs is substantial, especially in continental Europe. It is twice as high as in the Southern countries, and could be explained by the fact that in Mediterranean women, once they enter the labour market, appear to be more resilient compared to Continental women; career interruptions due to children do not translate into quitting work altogether. Overall, one in four women had no interruption whatsoever after the arrival of their (last) child.

The lowest rate of dropouts from work due to the birth of child are experienced in the Eastern European countries, ex-GDR, Sweden and, somehow surprisingly, Greece (probably due to dropping out at an earlier stage – e.g. at marriage). The highest dropout rates (over 20%) are recorded in Austria, Germany and the Netherlands, while a cluster including both continental (Belgium, France) and Southern countries (Spain, Italy) had the highest shares of women that did not interrupt their career at all when they had their children.

The duration of the interruption due to childbirth is presented in Fig. 12.3. As plainly illustrated, interruptions tend to be shorter in the low female participation countries (Southern European countries exhibiting a bipolar work pattern for women), followed by Denmark, Belgium, Poland and Sweden. The longest career interruptions occur in Switzerland and the Netherlands (Germany, France and Austria follow). Differences become more apparent when we look at groups of countries.

Table 12.1 Career interruptions due to children

Country	Never worked again (%)	Stopped working temporarily (%)	No interruption (%)	Women who worked at the time of 1st childbirth (N)
SE	7.5	86.1	6.3	791
DK	14.5	69.9	15.7	825
PL	5.7	59.1	35.2	722
CZ	0.4	94.4	5.3	990
DE-E	2.1	73.0	25.0	215
DE-W	29.5	47.0	23.4	603
NL	31.1	54.4	14.5	678
BE	18.3	34.3	47.3	889
FR	17.6	43.9	38.5	851
CH	17.7	54.5	27.8	451
AT	24.2	55.8	19.9	435
IT	17.9	43.9	38.2	519
ES	13.3	46.3	40.4	292
GR	6.0	69.8	24.2	534
Total	16.9	54.4	28.8	8.795

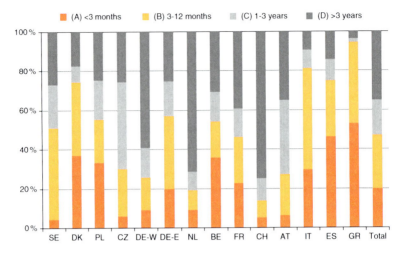

Fig. 12.3 Duration of interruption for females who stopped working temporarily because of a child, by country

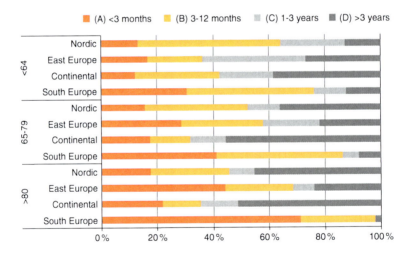

Fig. 12.4 Duration of interruption for females who stopped working temporarily because of a child, by cohort and country group

Almost 90% of working mothers in the Mediterranean countries interrupted their career for less than a year after the arrival of their (last) child, while about half of working mothers in Continental Europe experience work interruption longer than 3 years. This could reflect the lack of maternity protection (e.g. short maternity leaves) for working women at the time of childbearing of the SHARE group; they either had to leave altogether or get back very soon.

Figure 12.4 performs the same analysis by cohort and group of countries. It shows that there is a clear convergence for younger cohorts. The incidence of very short career interruptions (less than 3 months) declines everywhere in consecutive age cohorts. Interruptions up to 1 year increased only in the Nordic countries, declined in the East and the South, while remained fairly stable in the Continental countries. There seems to be a general trend for the period of absence to tend between 3 and 12 months, at the expense of both longer (>3 years) and shorter (<3 months) interruptions, presumably reflecting the maternity leave regulations. In general, the overall impression is one of convergence, with the Nordic countries in the lead.

12.4 Comparing Minorities: Two Exercises

Having established a number of different work–family patterns and their prevalence, we now examine women following atypical work–family patterns [exceptions to a broader rule in their country – for which see Brugiavini et al. (Chap. 13)]. This can be done by means of: First, comparing "minority women" with women following the same pattern but in contexts where this constitutes the rule ("majority women"). Second, try to establish whether "minority women" share more similarities with women following the same pattern elsewhere where this is the majority case or with the "majority" or "canon" women in their own context. The following points come out.

Exercise 1: The Full-Family-Model Women

- Full family (never worked) women tend to have more children than all other women, but the difference is larger when Full Family pattern is an exception.
- Where Full Family is the exception, the number of marriages is slightly higher, but when it is the rule it doesn't have any influence on the number of marriages.
- Full family women belong to poorer households, while their education is lower.
- Full family women are more likely to live in owned accommodation.
- Initial conditions matter in a variety of ways. Having experienced poverty as a child (few or no books at home) is more common for full-family women when they are the rule and considerably less when they are an exception (more opportunities and room for choice when more women work in a society?).

Exercise 2: The Full-Career-Model Women

- Women who have less than a full career are the same everywhere (regardless of context) in the sense that they have more children.
- Full career women tend to follow the general rule in their country, but have fewer numbers of marriages compared to the rest.
- Full career women tend to be more educated, but only when they are in the minority.
- Full career women tend to be richer, and this effect is stronger where they make up the majority trend (rule). Nevertheless, they have lower home-ownership rates.
- Where full career women are the rule, they have better self-perceived health scores, higher job and career satisfaction, fewer disappointments and fewer sacrifices for their job.

12.5 Explaining the Patterns: Does Social Policy and Employment Protection Matter?

The "naked eye" analysis so far has uncovered sea-changes in the patterns of female employment that have taken place in Europe over the life-span of the SHARE sample. To start uncovering relationships and the role of the policies, a multidimensional analysis must be the next step. The patterns we have seen, especially, among the older individuals signal that the crucial decision is taken early on in the life of women: whether to enter the labour market or not. Once having entered, most continue to a full career, though some drop off. To capture this pattern, our preliminary investigation employs a two stage analysis:

Firstly, the participation decision is modelled for the ever-entered group of women (i.e. those who have worked). Given that this is a decision adopted in their 20s, care is taken to include only those variables that would have been known at that time.

Secondly, the decision of how long to work is considered, conditional on having entered the labour force. The dependent variable is years of work to aged 50, in order to avoid bias as between women with completed working lives and those still working.

The structure of decisions is essentially recursive, where the participation and career length decisions are separated in time. Nevertheless, the fact that the group of women remaining in the labour market is essentially self-selected, creates a bias, implying that a simultaneous treatment of the two decisions is necessitated. Thus, a two stage Heckman type model is estimated (e.g. Maddala 1983), there being evidence of selection bias. Interestingly for the hypothesis that social policy matters, the selection bias evidence is much stronger once country groups are allowed.

In the *participation decision*, the effect we are trying to capture is the ease of entry in the labour market at the time when our sample was in their 20s. Thus the unemployment rate and the growth rate enter as proxies of labour market opportunities. The OECD Employment protection index is the earliest available – that of the mid-80s. Given that employment protection in Europe before that time was mostly greater than that, its value at the 80s can be taken as the minimum value for the appropriate decade. High employment protection can be expected to make labour force entry more difficult. The childhood relative well-being index is that of Lyberaki, Tinios and Georgiadis described in Chap. 2, and is a composite of the available indicators weighted according to their prevalence.

Table 12.2 shows a strong cohort effect, a positive effect of being an orphan, as of a large family size at childhood. The "mother at 22" variable (following Goldin 2006: 14) has a positive effect, though that should be evaluated with a large negative effect of the related "mother at first job" variable in the length of career equation. Interestingly, if country group dummies are introduced, it becomes negative, indicating the presence of an effect differentiated by country group (by a national difference not captured by the current specification). Relative childhood

Table 12.2 Determinants of participation decision in a simultaneous Heckman sample selection model (selected effects)

Dependent variable = Ever worked (i.e. >0 years of work)	Marginal effect	Standard error
Demographics		
Constant	4.8098**	0.2059
Mother when 22	0.0654*	0.0331
Orphan	−0.5148**	0.9597
Family size when child	−0.0749**	0.0127
Initial conditions		
Childhood relative well-being index: for each country ranges from 0 to1 (complete to no deprivation)	−1.4989**	0.1823
Occupation of breadwinner when 10: Legislator, senior official, manager, clerk	−0.1532**	0.0551
Occupation of breadwinner when 10: Elementary agricultural or fishery worker	−0.2251*	0.0632
Number of books when 10 (>10)	0.3964**	0.0393
Period of financial hardship up to 20	−0.4453**	0.0927
Primary education or lower	F	
Context variables when 20 (averages by cohort)		
GDP real growth rate	−1.3271	1.9794
Unemployment rate	−0.0105	0.0050
EPI index	−0.5434**	0.0308
Transition country(CZ; PL; GDR)	0.1097*	0.0715

Effects not reported: Age 65–80 and 80+ strongly negative; education strongly positive; foreign born, poor health when child – insignificant
Source: SHARE Wave 1, Wave 2 release 2.3.0; Wave 3, release 0
**,*: Significant at 1%, 5% respectively

well-being has a negative effect, signifying that participation in many cases may have been dictated by strained circumstances. This interpretation is strengthened by the negative effect of elementary occupation. For some women starting to work was an imposed necessity, for others an active choice.

Turning to context variables, an interesting pattern emerges: High national average unemployment in their 20s is associated with smaller entrance. High employment protection for those at work translates very strongly for problems to enter. Finally, though insignificant, high real GDP growth is associated with smaller entry probability. Adding country group dummies to this specification can be interpreted as allowing for the influence of social protection and labour protection "styles". This addition has the effect of almost eradicating the influence of household size, "mother at 22" and financial hardship. It also changes the sign on unemployment, reduces the influence of employment protection and makes the effect of growth large. These effects can be taken as evidence that those variables may have opposing influence in different policy settings; once the overall effect of country groups is allowed for, the within-group variation is able to exhibit itself. The weakening of public policy variables once the influence of large country groupings is allowed for can be taken as evidence that most of the public policy effect comes between country group variation.

Table 12.3 Determinants of career length in a simultaneous Heckman sample selection model (selected effects)

Dependent variable: Years of work to 50 given any work	Marginal effect	Standard error
Constant	28.5433**	0.4515
Demographics		
Number of children (1)	1.8167**	0.2319
Number of children (2)	0.8635**	0.2012
Number of children (3+)	0.1083	0.2184
Divorced	1.0048**	0.2232
Married when got first job	−3.8118**	0.2542
Mother when got first job	−6.9525**	0.2935
Ever left job because of ill health or disability	−1.3379**	0.1913
Occupational information		
Number of jobs	−0.6802**	0.0362
Ever been civil servant	0.6813**	0.2081
Got pension before the age of 50	1.4924**	0.2165
Historical Context variables when 40 (average by cohort)		
EPI 1980s (Employment Protection Index)	0.1498	0.1576
Unemployment by cohort	−0.0480	0.0244
Social protection on "family function" as (%) of GDP by age cohort	0.1923	0.1053
Minimum wage as % of average in 1970s, 1980s, 1990s by age cohort	−0.8307**	0.3024
Maternity leave length by age cohort	−0.0077	0.0066
Maternity leave replacement rate by age cohort	1.7939**	0.4001
Transition country(CZ; PL; GDR)	2.1350**	0.3444
Number of observations	12,125	
Censored observations (working)	2,074	

Effects not reported: Age 65–80 and 80+ strongly negative; education strongly positive. Occupation, industry, self-employed status, foreign born, poor health when child – insignificant
Source: SHARE Wave 1, Wave 2 release 2.3.0; Wave 3, release 0
**: Significant at 1%

To examine the *length of career decision* the dependent variable is defined as number of years worked until the age of 50. In this way the spurious correlation is avoided between age and years at work given that we have both women currently working and women already retired. A further advantage of this definition is to abstract from considerations relating to pensionable age, which are bound to introduce differentiations at the top end.

The continuous decision on career length appears in Table 12.3. *Children and Marriage:* Being married and being a mother at the time of labour force entry are both very important, subtracting 11 years from the predicted value. This confirms Goldin's (2006) observation for the US that being in the labour force before marriage and childbearing cements a permanently strong labour force attachment. The magnitude of the other children variables should be seen in this light: Beyond the first child, the marginal impact of an additional child is negative and increasing. *Education and health:* Given the cut-off at 50 the negative effect of education is due to later entry. Poor health is important mainly if the problem was sufficiently serious to necessitate leaving a job – reentry presumably is then harder.

Occupation: The frequency of changing jobs leads to a lower expected length (reentry problems). Later entry presumably accounts for shorter careers in public administration (corrected by a positive sign for being a civil servant). Owning a business has an effect on length of career, as is being eligible for a pension before 50.

Context variables: A high minimum wage relative to the average leads to reductions in careers, as returns to the labour market after an interval of absence are more difficult. Social protection expenditure on family policies and the replacement rate of maternity allowance have an influence. Unemployment has a negative effect (as it did in participation), implying that high unemployment prevented labour entry and reduced careers. The EPI index appears only to affect participation and to have no effect on career length.

However, once the same specification is run with country group dummies, the influence of social policy context variables is completely transformed. The key differences must be due to the differences of the "Mediterranean welfare states" (to follow Ferrera rather than Esping-Andersen). Once the southern European factor is allowed for, virtually all context variables become significant and have the expected signs, meaning that they explain differences within groups, whereas differences between groups must be due to more diffuse systemic differences which interact with our simple variables. For instance, the EPI index is large and negative, while social protection on the family function becomes large and positive. Indeed, in the typologies of Welfare states, the Mediterranean state is supposed to stand out by placing all emphasis on pensions and none at all on the family function. During the working life of the SHARELIFE sample, in the Mediterranean both social protection family policy would have been absent, while employment protection would not be extended to women.

Given that most of our context variables essentially capture social policy effort, the transformation of the effects once a generalized "Southern" effect is allowed for, implies that the same effort in different parts of Europe had different effects. This can be taken as a strong indication that – in the period when our sample were still young – the workings of the welfare systems and the way those related to the economy were to a large extent distinct – at least as between North and South (Table 12.4).

12.6 Conclusions

The 50 years encompassed in the lives of women in the SHARELIFE sample capture the periods of development, apogee and consolidation of distinct "worlds of welfare capitalism" into what many call the "European Social Model". Has our analysis shone light of this process?

- The working lives initially followed a polarized pattern which is becoming less so with time. Rather than two distinct groups, younger cohorts exhibit more women with adaptive careers, leaving and reentering the labour market.

Table 12.4 Context variables in the Heckman regression once country groups are distinguished (selected coefficients only)

Dependent variable = Years of work to 50 ∣ work>0	Marginal effect	Standard error
Constant	29.6813**	0.9259
Historical Context variables when 40 (avg by cohort)		
EPI 1980s (Employment Protection Index)	−0.6731**	0.1814
Unemployment by cohort	−0.0573*	0.0263
Social protection on "family function" as (%) of GDP by cohort	1.4290**	0.1996
Minimum wage as % of average by age cohort	−0.1361	0.3481
Maternity leave length by age cohort	−0.0403**	0.0087
Maternity leave replacement rate by age cohort	1.5538**	0.4080
Continental	−0.2633	0.3156
South	2.0719**	0.7418
Transition	3.0010**	0.3981
Selection equation == Ever worked Historical Context variables when 20 (avg by cohort)		
EPI 1980s (Employment Protection Index)	−0.0734	0.0400
Unemployment by cohort and 1960s, 1970s, 1980s	0.0097	0.0053
Growth rate by cohort and 1970s, 1980s, 1990s	−3.9046*	1.9834
Continental	−0.9602**	0.0922
South	−1.8238**	0.0953
Transition	−0.5591**	0.1141

Source: SHARE Wave 1, Wave 2 release 2.3.0; Wave 3, release 0
**,*: Significant at 1%, 5% respectively

- This process is visible everywhere, but it is very uneven still in its geographical and social spread.
- The econometric evidence finds some evidence for convergence. Social policy matters more for the length of career, rather than for participation – which was taken earlier and on the basis of the situation pertaining before the 1980s.
- There appear two large fissures in Europe: one regarding the transition countries, and another regarding the Mediterranean. Indeed social policy parameters seem to change their meaning and significance once we allow for a generalized "Mediterranean" effect.

Do policies matter? Our verdict is "undoubtedly yes". However, the same policies may produce very different outcomes, while similar outcomes may correspond to very different policies (Daly 2002). It is interesting to hypothesize on "functional equivalents", i.e. factors Y that in country A produced results brought about by policy X in country B. In this case, Y is a de facto functional equivalent of X. Lack of public social infrastructure may be compensated by the market for such services or even by quiet grannies. The European welfare state encompasses the formal social policy apparatus in the North, and an informal family-based support system in the South.

The big story the researchers should not lose out on is the steady but sure convergence of family and work patterns. This convergence still leaves much ground uncovered. Much of the differentiation in older women is the result of older discrimination and cumulated inequities which necessitate special attention.

References

Bettio, F., & Villa, P. (1998). A Mediterranean perspective on the breakdown of the relationship between participation and fertility. *Cambridge Journal of Economics, 22*, 137–171.

Boeri, T., (2003). *Women on the 'low equilibrium'*, mimeo.

Boeri, T., Del Boca, D., & Pissarides, Ch (Eds.). (2005). *Women at work: an economic perspective* (Fondazione Rodolfo DeBenedetti). Oxford: Oxford University Press.

Daly, M. (2002). A fine balance: Women's labor market participation in international comparison. In F. W. Scharpf & V. A. Schmidt (Eds.), *Welfare and work in the open economy* (Vol. II, pp. 467–510). Oxford: Oxford University Press.

Goldin, C. (2006). The quiet revolution that transformed women's employment, education and family. *The American Economic Review, 96*(2), 1–21.

Hakim, K. (2000). *Work-lifestyle choices in the 21st century: Preference theory*. Oxford: Oxford University Press.

Hakim, K. (2004). *Key issues in women's work: Female diversity and the polarisation of women's employment*. London: Glasshouse Press.

Lewis, J. (2001). The decline of the male-breadwinner model: The implications for work and care. *Social Politics, 8*(2), 152–170.

Lewis, J., Campbell, M., & Huerta, C. (2008). Patterns of paid and unpaid work in Western Europe: Gender, commodification, preferences and the implications for policy. *Journal of European Social Policy, 18*(21), 21–37.

Maddala, G. S. (1983). *Limited dependent and qualitative variables in Econometrics*. Cambridge: CUP.

Chapter 13
Maternity and Labour Market Outcome: Short and Long Term Effects

Agar Brugiavini, Giacomo Pasini, and Elisabetta Trevisan

13.1 Maternity and Labour Market Outcomes

Retirement patterns, as well as continuity and length of work histories, are strongly influenced by the events over the life cycle. This is particularly true for women. Gender differences in work-careers and the role of women within the family usually lead to fewer pension rights and lower retirement income for women than for men. In particular, maternity is likely to be one of the major drivers of gender differences on life time economic outcomes such as labour force participation, differential productivity and wages and eventually retirement income.

The effect of motherhood on women's labour supply has been a long-standing focus of economic research seeking to explain the rise in the labour force participation of women over the past decades, together with the decline in fertility rate; it has also been the cornerstone of public policy regarding the efficient design of parental leave and benefit (Troske and Voicu 2009). The research on this topic focuses on the effect of timing and spacing of births to explain the decrease in fertility characterizing OECD countries; a few studies look instead at the effect of timing and spacing of births on women's labour market outcome (Gustafsson 2001). One of the main results is that timing and spacing between births matter substantially for labour market participation: women who have the first child later in life exhibit a lower probability of dropping out of the labour force and lower negative effects on wages compared to other women. Moreover, the drop-out effect induced by maternity increases with the number of children. As for the long-term effects of maternity, very little is known and the evidence is very scanty.

The aim of this paper is to fill this gap by analyzing the long term effects of childbearing, i.e. the effect of motherhood on pension income at retirement, given the labour market participation of women at childbirth. Since labour market attachment is higher for younger generations, it is relevant for policy makers to look at the

---

A. Brugiavini (✉), G. Pasini, and E. Trevisan
Dipartimento di Scienze Economiche, Università Ca' Foscari Venice, Cannaregio, 873, 30121 Venice, Italy
e-mail: brugiavi@unive.it; giacomo.pasini@unive.it; trevisel@unive.it

behaviour of women who want to work excluding those who plan a "family-life" (see also Chap. 12). SHARELIFE is particularly suitable for this analysis since it contains complete life time histories, including all the employment and maternity episodes experienced by European women currently aged 50 and over. Moreover, details on maternity leave provisions and other institutional features of the SHARE countries are collected and provided together with the survey data. These institutional features allow us to investigate if and how the presence of maternity benefits affects the labour market participation decisions of women after childbirth and, consequently, the impact of pension income at retirement.

13.2 Maternity Leave Across Europe

The existence of maternity leave provisions at the time of motherhood is likely to influence the labour market participation of women and pension income at retirement. The main characteristics of maternity leave are the duration and the amount of the benefit, the latter usually expressed as percentage of the wage. Table 13.1 summarizes these variables for each country.

There is variability between countries in terms of maternity "protection". In all European countries, with the exception of Austria and Czech Republic, maternity leave provisions have been introduced in 1970 and have gone through several changes. Sweden, Denmark and Italy are characterized by a longer duration of the maternity leave with respect to other countries. However, all countries show a trend towards the increase of duration of maternity benefits in the last decades.

Table 13.1 Maternity leave provisions across Europe

Country	Year of introduction	Duration (weeks)	Benefit (%)
Sweden	1970	26–64	48–90
Denmark	1970	14–30	88–100
Germany	1970	14	100
Netherlands	1970	12–16	100
Belgium	1970	14–15	60–77
France	1970	14–16	50–100
Switzerland	1970	10–16	100
Austria	1950	12–16	100
Italy	1970	17–21.5	80
Spain	1970	12–16	75–100
Greece	1970	12–16	50
Poland	1974	16	100
Czech Republic	1961	28	69

Note: Benefit is expressed as a percentage of wages in the manufacturing sector. Maternity leave provisions information are not available for East Germany separately from West Germany. The two numbers for the duration and the benefit columns indicate respectively the lowest and the highest number of weeks and benefit level fixed by law during the years

Countries with shorter durations compensate with generous benefits, varying between 80 and 100% of the wage. Greece, Belgium and the Czech Republic are exceptions.

13.3 Short Term Effects

One important economic effect of childbearing is a change in labour market participation of women after childbirth. Hence we look at the subsample of women who were working at the time of motherhood (for any child), which is clearly a selected group of the population. Figure 13.1 shows that the labour market participation rate of women at the time of childbirth is indeed very heterogeneous across countries.

In some countries, such as Italy and Spain, the fraction of women working at the time of childbirth is very low (about 30%). In most of the other countries the labour market participation rate of women is above 50%, with peaks of 90% for Czech Republic and East Germany.

In SHARELIFE, for each maternity episode, it is asked whether the respondent continued working without interruption, stopped temporarily her job (maternity leave) or left the labour market and never worked again. The average drop-out rate from the labour market at childbearing by country and number of children is shown in Table 13.2. The table shows that the choice of whether to continue working for women, in case of motherhood, is influenced by the number of children. We observe an increasing frequency of exits from the labour market as the number of children increases for almost all countries. Table 13.2 shows also a high variability between countries, which is likely to be related to country-specific cultural differences as well as to differences in maternity leave provisions.

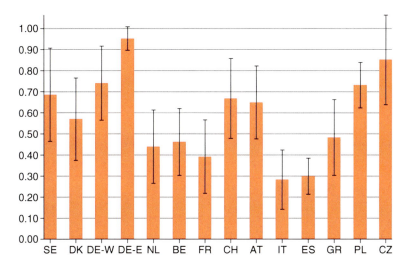

Fig. 13.1 Fraction of women working at the time of childbearing by country

Table 13.2 Drop-out rate at childbearing by country and number of children

Country	One child	Two children	Three children	Four children
Sweden	0.06	0.06	0.07	0.10
Denmark	0.22	0.12	0.19	0.32
West Germany	0.42	0.24	0.29	0.38
East Germany	0.35	0.28	0.48	0.24
Netherlands	0:26	0.12	0.17	0.06
Belgium	0.51	0.57	0.50	0.49
France	0.48	0.51	0.54	0.49
Switzerland	0.34	0.23	0.28	0.33
Austria	0.22	0.25	0.27	0.35
Italy	0.39	0.43	0.41	0.45
Spain	0.51	0.33	0.44	0.27
Greece	0.22	0.26	0.15	0.30
Poland	0.31	0.31	0.42	0.57
Czech Republic	0.04	0.06	0.11	0.09

Note: The drop-out rate is calculated as the ratio between the number of women who were working at the time of childbirth and never worked again and the total number of women who were working at the time of childbirth (by number of children)

Figure 13.2 shows country averages of labour market consequences of childbearing, conditional on the number of children and on the existence of maternity benefit at the time of childbirth.

Comparing labour force participation in the absence (left panel) and presence (right panel) of maternity benefits at the birth of the oldest child, a high variability between countries emerges. The puzzling result is that in some countries the existence of maternity benefits seems to increase the probability of dropping out of the labour market. It should be noted that, in this particular case, the sample size in some countries may be small as we restrict the attention to women with two children.

The introduction of maternity leave provisions aimed at mitigating the income drop of mothers and at providing incentives to stay in the labour market. The descriptive evidence is mixed: the overall effect of these policies on labour market participation depends on the cohort of the mother, as well as on the number of children at the moment of childbearing. Even after controlling for these characteristics we obtain a large cross-country heterogeneity in labour force participation. This seems to indicate that the effectiveness of public policies for maternity protection does not depend solely on length and generosity of those benefits, but also on individual preferences, cultural traits, as well as other transfers in money and in kind at childbearing, such as public day-care provisions.

13.4 Long Term Effect

The number and timing of children could also have long term effects. In particular, pension provisions are typically related to working life history: the number and length of employment interruptions of mothers could affect their social security (pension) income at retirement (Boeri and Brugiavini 2009).

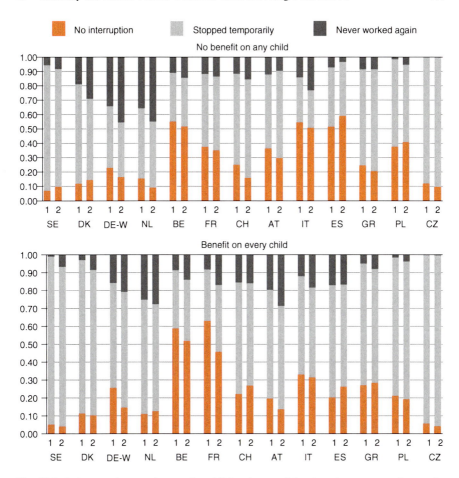

Fig. 13.2 Labour market attachment after childbearing conditional on the presence of maternity benefits (women with two children). *Note*: In the graphs women with two children are considered, for each child the labour market outcome is shown. Czech Republic and Austria are not present in the *left panel* (without maternity benefit) because all women having two children are covered by maternity leave provisions

In order to explore such a long term effect, we first look at the relation between the number of children, the labour market participation decision at childbirth and the social security (pension) replacement rate. We define the replacement rate as the ratio of the first pension benefits received after retirement and the last wage received. Figure 13.3 shows the relationship between the replacement rate and the number of children at the aggregate level.

What emerges is a relatively high variability in fertility rates across countries, compared with the dispersion of actual replacement rates. As a result, there is a small negative correlation between the two, which is not statistically significant.

The aim of this analysis is to identify the effect of the existence and generosity of maternity leave programs on the pension income at retirement. Thus, we focus on

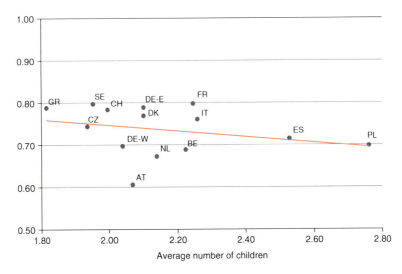

Fig. 13.3 Average pension replacement rate by country and number of children

those women who had at least one child and who were working at the time of the first childbirth. More specifically, starting from a sample of 15,544 women, 15,523 have at least one child and among them 8,963 were working at the time of first childbirth. About 44% of these women (3,986) are retired, while the majority of them is still working, disabled, unemployed or out of the labour force for other reason.

We now focus on the effect of maternity leave provisions on pension income at retirement at individual level. This effect is captured by three variables: a dummy variable accounting for the existence of maternity leave provisions at the time of first childbirth, a variable describing the benefit received during maternity leave as a percentage of the wage and a variable indicating the length of maternity leave in weeks. If maternity leave provisions are introduced after the first motherhood, women may change their fertility decisions according to these policy decisions. Modelling the interplay between individual fertility choices and the timing of the introduction of maternity leave benefits goes beyond the scope of this paper, thus we consider the existence and the characteristics of maternity leave provisions only at the time of first childbirth. Moreover, we include simultaneously country dummies among the regressors besides maternity leave indicators: while country dummies capture country-specific effects, the maternity leave indicators capture variations both across countries and over time.

In Table 13.3 we present the results of the regression analysis for the log of social security pension income when controlling, in turn, for the existence of maternity leave (column 1), the benefit level (column 2) and the length of maternity leave (column 3). The length and generosity of maternity leave provisions have a negative and significant effect on pension income at retirement. Long spells due to maternity leaves might induce women to stay longer out of labour force and longer non-employment spells could have worsened the labour market opportunities after

Table 13.3 Regression analysis of the social security pension income (log)

Variables	(1)	(2)	(3)	(4)
Maternity leave provision:				
Existence	0.204***			0.794***
	(0.039)			(0.217)
Maternity benefit replacement rate		−0.002***		−0.006**
		(0.000)		(0.002)
Length			−0.010***	−0.005
			(0.003)	(0.005)
Stopped temporarily	−0.094**	−0.095**	−0.093**	−0.091**
	(0.039)	(0.039)	(0.039)	(0.039)
Drop-out	−0.400***	−0.401***	−0.409***	−0.402***
	(0.067)	(0.067)	(0.067)	(0.067)
Number of children	−0.033**	−0.033**	−0.038***	−0.035**
	(0.019)	(0.014)	(0.013)	(0.014)
F-test				F(3.2744) = 11.90
				Prob>F = 0.0000
N. Obs.	3,986	3,986	3,986	3,986

Note: *, **, *** stand for 90%, 95% and 99% level of significance respectively. The dependent variable is the log of first pension benefit at retirement, Existence is a dummy variable taking value 1 if there were maternity leave provisions at the time of first childbirth and 0 otherwise; Maternity benefit replacement rate is the institutional maternity benefit level (as % of the wage) at the time of first childbirth; Length is the institutional length of maternity leave at the time of first childbirth. In all models we control for years of education, job characteristics, and country dummies. The amounts of the benefits at retirement have been corrected for within countries differences in currencies and have been converted in Euro (exchange rate 2001)

re-entering the job market, thus reducing wages and benefits compared to women with shorter interruption (this is in line with Troske and Voicu 2009). Considering the three variables together (column 4), the existence of maternity benefit has a positive and significant effect, while the level of the benefit has a negative and significant effect (the three variables are jointly significant). Moreover, the pension income at retirement is lower both for women leaving temporarily their job and for women dropping out from labour market compared to women without interruptions in their careers. There are no major differences across countries.

Hence our investigation suggests that, by and large, time spent out of the labour market and generous maternity benefits could have a negative effect on the pension income at retirement.

However, the fertility decision of women is likely to be endogenous. In order to deal with the simultaneity issue, we estimate the effect of maternity leave provision on the pension income at retirement using an instrumental variable approach, where the number of biological brothers and sisters of the mother at the age of 10 is used as instrument for the number of children. The idea is that the fertility history in the household in which the woman grew, which is reflected in the number of brothers and sisters, influences her fertility decision, but it is not related to her labour market participation. More precisely, the number of brothers and sisters at the age of 10 could influence the labour market participation of women at very young ages (women with more siblings might be pushed to go to work earlier), but it is unlikely to have any direct effect on labour market participation choices later in life. In our analysis we consider women who were working at the time of childbearing and we look at the labour market participation after childbirth, thus, on the basis of the above arguments, we can assume that our instrument is still valid. Table 13.4 shows the estimation results.

As in Table 13.3, the results of Table 13.4 show that women who left their job temporarily and women dropping out from the labour market after childbirth have a significantly lower pension income at retirement than women who did not leave their jobs. There are no significant differences across countries.

The IV estimates are very much in line with the ones deriving from the OLS regression. More specifically Table 13.4 shows (columns (1) to (3)) that, when the variables related to the maternity leave provisions are included separately, the existence of maternity leave provisions has a positive and significant effect, while length and generosity of maternity benefit have a (negative) significant effect on the social security pension benefit. When the "maternity variables" are included together (column 4), while the existence of maternity leave provisions has a strong and positive effect, the level of the maternity benefit reduces the social security pension income at retirement and the length of the benefits is not statistically significant.

Table 13.4 Instrumental variable regression of the pension income (log)

Variables	(1)	(2)	(3)	(4)
Maternity leave provision:				
Existence	0.183***			0.822***
	(0.060)			(0.231)
Maternity benefit replacement rate		−0.002**		−0.007**
		(0.000)		(0.002)
Length			−0.009***	−0.005
			(0.004)	(0.005)
Stopped temporarily	−0.102**	−0.103**	−0.101**	−0.098**
	(0.044)	(0.043)	(0.044)	(0.043)
Drop-out	−0.392***	−0.393***	−0.402***	−0.394***
	(0.069)	−(0.069)	−(0.069)	(0.069)
Number of children	−0.078	−0.079	−0.080	−0.083
	(0.107)	(0.107)	(0.107)	(0.107)
F-test				F(3.2740) = 8.64
				P>F = 0.000
N. Obs.	3,977	3,977	3,977	3,977

Note: *, **, *** indicate 90%, 95% and 99% level of significance respectively. In the regressions the dependent variable is the log of first pension benefit at retirement, Existence is a dummy variable taking value 1 if there were maternity leave provisions at the time of first childbirth and 0 otherwise; Maternity benefit replacement rate is the institutional maternity benefit level (as % of the wage) at the time of first childbirth; Length is the institutional length of maternity leave at the time of first childbirth. In all models we control for a set of variables, such as years of education, job characteristics, country dummies. The amounts of the benefits at retirement have been corrected for within countries differences in currencies and have been converted in Euro (exchange rate 2001)

We can conclude that the existence of maternity benefits improved long term economic conditions for mothers, but the generosity (replacement rate of the maternity benefit) of the maternity benefit had a negative effect on pension income at retirement.

13.5 Conclusion

The number and timing of children are likely to affect labour market participation of women (what we called short time effects) and, through the induced discontinuity of work careers, they also affect retirement income.

In this paper we first describe the labour market attachment of women at childbirth, then we analyze the long term effect of childbearing, i.e., the pension income at retirement, focusing on the effect of the existence and characteristics of the maternity leave arrangements provided in each country.

The main results in terms of "short-term" and "long-term" economic outcomes can be summarised as follows:

- The "exit from the labour market" effect due to childbearing is increasing in the number of children.
- There is a high variability in the pattern of participation across countries related to differences in maternity leave provisions; however, institutional features are not the only determinants of the interaction between fertility and labour force participation.
- The existence of maternity leave provisions has a positive effect on the social security pension benefit at retirement.
- The generosity of maternity benefits reduces the social security pension income.
- Results hold even after controlling for the endogeneity of fertility decisions.

Overall, both for short-term labour market outcomes and for the social security pension income, the introduction of maternity benefits, aimed at mitigating the reduction in income of mothers and at providing incentives not to leave the labour market, did lead to an improvement in the economic conditions of mothers, but the characteristics of the maternity leave provisions such as the level of the benefit may reduces their beneficial effect.

References

Boeri, T., & Brugiavini, A. (2009). Pension reforms and women retirement plans. *Journal of Population Ageing, 1*, 7–30.

Gustafsson, S. (2001). Optimal age at motherhood. Theoretical and empirical considerations on postponement of maternity in Europe. *Journal of Political Economy, 85*, 225–217.

Troske, K.R., & Voicu, A. (2009). *The effect of the timing and spacing of births on the level of labor market involvement of married women.* IZA Discussion Paper, N. 4417.

Chapter 14
Reproductive History and Retirement: Gender Differences and Variations Across Welfare States

Karsten Hank and Julie M. Korbmacher

14.1 Multiple Perspectives on the Fertility–Employment Nexus in Later Life

The association between women's fertility and employment has received considerable attention in the social science literature, which documents significant variation in the observed correlations across *time* and between *welfare states* (Ahn and Mira 2002). Moreover, the employment–fertility nexus has been shown to vary by *gender*: mothers might suffer from limited opportunities for paid employment under the same institutional regime that allows for positive income effects of fatherhood, for example.

Parenthood and its immediate consequences for labour market outcomes are also likely to exhibit sustained influences on employment patterns over the *family life course*, particularly for women (Chap. 13). Even though children need not necessarily reduce one's total years in the labour force, they may create discontinuities in employment careers eventually affecting individuals' pension receipt. Thus "the retirement process is appropriately viewed as temporarily embedded in current incentive–disincentive structures that mediate retirement decision-making and in long-term family relations that constitute the joint role pathways of couples through work and family domains" (O'Rand et al. 1992: 82; italics in the original).

Only few studies have yet investigated the long-term relationship between individuals' reproductive history and retirement, though. The limited evidence collected so far tends to suggest that having children might actually delay women's exit from the work force (Hank 2004). These and related findings have been interpreted as reflections of *gendered role patterns* developed across the family life course (Henretta et al. 1993;

K. Hank (✉)
Institute of Sociology, University of Cologne, Greinstr. 2, 50939 Cologne, Germany
e-mail: hank@wiso.uni-koeln.de

J.M. Korbmacher
Mannheim Research Institute for the Economics of Aging (MEA), L13, 17, 68131 Mannheim, Germany
e-mail: korbmacher@mea.uni-mannheim.de

Pienta et al. 1994), but they are also likely to be affected by *welfare state* policies and institutions – which are not gender neutral either (Daly and Rake 2003).

Students of the interplay between childbearing and later life labour force exits should thus ideally take a cross-national and gendered perspective. With its broad set of retrospective life history information from respondents in 13 Continental European countries, SHARELIFE offers unique opportunities for researchers to conduct such kind of analysis, allowing us to add to the existing literature by providing initial answers to three important questions in particular:

- Are there systematic *gender* differences in the association between reproductive history and retirement?
- Are longstanding differences in fertility, employment, pensions, and *welfare regimes* reflected in cross-national differences in the association between individuals' reproductive history and retirement? One might assume, for example, that institutional contexts *fostering* younger mothers' employment (e.g. Sweden) and such *prohibiting* women's labour force participation after childbirth (e.g. Italy) also exhibit such patterns if older women's employment is considered.
- Are there *cohort* differences in the association between reproductive history and retirement?

14.2 Reproductive and Employment Histories in SHARELIFE

In our study, we track the annual employment histories of 16,454 ever married men and women aged 50 or older who reported at least one episode of gainful employment. These individuals contribute a total of 156,033 person-years of observation to the analysis. Running separate regressions for men and women, we estimate discrete-time logit models to analyse the individual's entry into retirement.

For our dependent variable "retirement" there is no exclusive definition. It is operationalised here as an exit from the labour force after age 50 which is not followed by a re-entry into paid employment in any subsequent time period (note that observations are censored at age 80 or at the time of the SHARELIFE interview); see Blanchet et al. (2005) for an overview of different pathways to retirement in SHARE countries.

Our main *explanatory* variables are three indicators of individuals' fertility: first, the number of children ever born (with an average of 2.1 in our sample); second, a dummy variable indicating whether the individual ever had a child (note that 8–9% of the respondents in our sample remained childless); and third, binary indicators of an early (late, respectively) first birth, where the threshold was set at age 24 for women and age 27 for men (i.e. the median of the distribution for the pooled sample). Such detailed new information on men's and women's *reproductive history* substantially increases SHARE's research potential in this domain (see Martínez-Granado and Mira 2005, for an analysis of fertility using data from SHARE's Wave 1).

Further *control* variables include individuals' age, level of education, work experience (years in the labour force and number of jobs till the age of 50), marital status as well as indicators of self-employment and homeownership (Hank 2004). To account for possible cohort differences in the relationship between fertility and retirement, we distinguish between respondents born up until 1940 (who account for roughly one third of our sample) and those born later.

For our *regionalized* analyses, we use Esping-Andersen's (1990) initial welfare state typology to group the countries represented in SHARELIFE into four clusters which we label as Social-democratic (DK, NL, SE), Conservative (AT, BE, CH, DE, FR), Mediterranean (ES, GR, IT), and Post-communist (CZ, PL).

14.3 Parenthood, Number of Children, Timing of Fertility: What Determines Older Individuals' Exit from the Labour Market?

Before turning to the discussion of the main explanatory variables, which also considers cohort and welfare regime interactions (see Table 14.2), we first describe the results for all other control variables in an initial model for the pooled SHARE-LIFE sample (Table 14.1). Most of these findings are consistent with previous research (Hank 2004) and very similar for men and women.

To begin with, the odds of entering retirement increases significantly with individuals' age, sky-rocketing after age 59 and yet again at ages 64 and over. Moreover, members of the post-1940 cohorts retire later than those born before World War II. With regard to the role of a woman's marital status, there is clear indication that married women enter retirement earlier than their unmarried counterparts, whereas the respective coefficient remains insignificant for men. This finding might be interpreted as evidence supporting the "joint retirement hypothesis" (Henretta et al. 1993), which states that (on average older) husbands and (on average younger) wives tend to retire at the same time, resulting in earlier retirement among married women.

Women with a high educational degree tend to stay in the labour force longer than their low qualified counterparts. The number of years in employment till the age of 50 significantly influences the individual's decision to retire. A longer employment record results in earlier eligibility for old-age pensions and tends to speed up the transition to retirement. Moreover, those who were self-employed in their last reported job exhibit markedly lower risks of leaving the labour force than all others. Career interruptions (indicated by the number of employment spells) as well as homeownership, however, do not bear any association with the timing of retirement.

Table 14.2 summarises the results for different indicators of men's and women's reproductive history. In the initial model (also exhibited in Table 14.1) we operationalised individuals' reproductive history by their number of children. The risk

Table 14.1 Discrete-time logistic regression results for older women's and men's exit from the labour force, odds ratios (standard errors) – pooled initial model

Demographics	Women	Men
Age 50–53a	1.00	1.00
Age 54–57	2.92***	2.52***
	(0.136)	(0.124)
Age 58–59	4.20***	4.71***
	(0.252)	(0.259)
Age 60–61	14.55***	15.83***
	(0.819)	(0.806)
Age 62–63	13.06***	13.47***
	(0.916)	(0.804)
Age 64+	19.12***	29.89***
	(1.310)	(1.704)
Born 1940 or later	0.82***	0.93**
	(0.028)	(0.028)
Married	1.12***	1.08
	(0.046)	(0.057)
Education and work		
Low education	1.00	1.00
Medium education	0.91**	1.05
	(0.035)	(0.039)
High education	0.70***	0.96
	(0.034)	(0.044)
Years in labour force at age 50	1.02***	1.07***
	(0.002)	(0.004)
Number of jobs at age 50	1.00	1.00
	(0.010)	(0.009)
Last job self-employed	0.42***	0.37***
	(0.022)	(0.015)
Homeowner	0.95	1.03
	(0.034)	(0.034)
Reproductive history (also see Table 14.2)		
Number of children	1.00	0.98*
	(0.013)	(0.011)
Constant	0.02***	0.00***
	(0.002)	(0.000)
Pseudo-R2	0.13	0.16
Observations	64,528	91,446

Significance: *** = 1%; ** = 5%; * = 10%

of entering retirement among women born 1940 or later tends to slightly increase with a growing number of children (Table 14.2a). This finding seems to be driven mainly by women living under social-democratic and post-communist regimes. The reverse association is observed among men (Table 14.2e). An alternative specification, based on an indicator of whether the respondent ever had a child, also suggests that, independent of family size, in cohorts born post-1940 fatherhood (at least in social-democratic countries) bears a negative correlation with retirement (Table 14.2f), whereas motherhood contributes to an earlier exit from the labour force (Table 14.2b). Moreover, mothers who experienced an early first birth (particularly if they live in

14 Reproductive History and Retirement: Gender Differences and Variations

Table 14.2 Discrete-time logistic regression results for older women's and men's exit from the labour force, odds ratios (standard errors) – Interaction of reproductive history indicators with cohort and welfare state regime

Women	(a) Number of children			(b) Ever had a child		
Welfare regimes	All	Cohort <1940	>1940	All	Cohort <1940	>1940
Social-democratic	1.07**	1.06	1.11**	1.09	1.05	1.18
	(0.03)	(0.04)	(0.05)	(0.13)	(0.18)	(0.20)
Conservative	1.00	0.97	1.02	1.06	0.85	1.19
	(0.02)	(0.03)	(0.03)	(0.10)	(0.12)	(0.16)
Mediterranean	0.96	0.96	0.97	1.17	1.06	1.29
	(0.03)	(0.04)	(0.05)	(0.16)	(0.20)	(0.26)
Post-communist	1.04	1.00	1.11**	1.04	0.85	1.18
	(0.03)	(0.05)	(0.05)	(0.17)	(0.22)	(0.26)
All	1.00	0.98	1.04*	1.09	0.95	1.21**
	(0.01)	(0.02)	(0.02)	(0.07)	(0.08)	(0.10)

Women	(c) Early first birth			(d) Late first birth		
Welfare regimes	All	Cohort <1940	>1940	All	Cohort <1940	>1940
Social-democratic	1.22	1.25	1.21	1.07	1.11	1.05
	(0.15)	(0.22)	(0.21)	(0.13)	(0.19)	(0.19)
Conservative	1.16	0.92	1.33**	0.98	0.88	1.00
	(0.11)	(0.13)	(0.18)	(0.10)	(0.13)	(0.14)
Mediterranean	1.12	0.99	1.27	1.20	1.12	1.27
	(0.16)	(0.20)	(0.27)	(0.17)	(0.22)	(0.26)
Post-communist	1.17	1.03	1.27	1.00	1.07	0.99
	(0.18)	(0.25)	(0.27)	(0.17)	(0.28)	(0.22)
All	1.17***	1.04	1.29***	1.06	1.03	1.08
	(0.07)	(0.09)	(0.11)	(0.07)	(0.09)	(0.09)

Men	(e) Number of children			(f) Ever had a child		
Welfare regimes	All	Cohort <1940	>1940	All	Cohort <1940	>1940
Social-democratic	0.97	0.98	0.94	0.86	0.97	0.73**
	(0.02)	(0.03)	(0.04)	(0.09)	(0.15)	(0.11)
Conservative	0.98	0.99	0.96	1.06	1.11	0.98
	(0.02)	(0.02)	(0.03)	(0.09)	(0.13)	(0.13)
Mediterranean	0.97	0.97	0.97	0.96	0.88	1.01
	(0.02)	(0.03)	(0.04)	(0.10)	(0.13)	(0.15)
Post-communist	1.01	1.07	0.97	1.22	1.23	1.24
	(0.03)	(0.05)	(0.05)	(0.23)	(0.32)	(0.35)
All	0.98*	0.99	0.97*	1.00	1.04	0.95
	(0.01)	(0.01)	(0.02)	(0.05)	(0.08)	(0.07)

Men	(g) Early first birth			(h) Late first birth		
Welfare regimes	All	Cohort <1940	>1940	All	Cohort <1940	>1940
Social-democratic	0.93	1.02	0.81	0.78**	0.94	0.62***
	(0.10)	(0.15)	(0.13)	(0.09)	(0.14)	(0.10)
Conservative	1.12	1.19	1.00	1.01	1.04	0.94
	(0.10)	(0.14)	(0.13)	(0.09)	(0.13)	(0.13)
Mediterranean	1.05	0.96	1.09	0.92	0.89	0.95
	(0.11)	(0.15)	(0.17)	(0.10)	(0.13)	(0.14)

(continued)

Table 14.2 (continued)

Men	(g) Early first birth			(h) Late first birth		
Post-communist	1.24	1.28	1.23	1.25	1.26	1.22
	(0.22)	(0.32)	(0.33)	(0.23)	(0.33)	(0.34)
All	1.05	1.10	0.99	0.94	1.00	0.88
	(0.06)	(0.08)	(0.08)	(0.05)	(0.08)	(0.07)

Note: Social-democratic: DK, NL, SE; Conservative: AT, BE, CH, DE, FR; Mediterranean: ES, GR, IT; Post-communist: CZ, PL. All control variables used in the initial model (cf. Table 14.2) are also included here. Significance: *** = 1%; ** = 5%; * = 10%.

a conservative country) retire earlier than their childless counterparts (Table 14.2c), a result which – again – appears to be driven mainly by "younger" cohorts of women. Conversely, "late" fathers living under a social-democratic regime are shown to retire later than childless men (Table 14.2h).

14.4 Summary

The main findings of our initial assessment of the association between men's and women's reproductive history and retirement in Continental Europe can be summarised as follows:

- Different from previous research for Germany (Hank 2004), our findings based on SHARELIFE suggest that mothers (just as married women) are more likely than childless women to exit the labour force early, whereas fathers tend to retire later than other men. These findings might, in general, be interpreted as a reflection of a weaker labour force attachment among mothers and a greater responsibility of fathers as male breadwinners lasting well into individuals' late career phase.
- However, the association between childbearing and earlier retirement appears to be particularly strong among women living under a social-democratic or post-communist welfare state regime, that is, in countries exhibiting relatively high levels of female labour force participation. This might indicate that mothers in conservative and Mediterranean countries who re-entered the labour market after childbirth may be a selective group characterised by an above average attachment to employment. To investigate further the role of specific welfare state interventions (particularly such supporting mothers' employment or granting them special pension entitlements) is beyond the scope of this paper and therefore remains an important task for future research.
- There is some indication for a closer relationship between individuals' reproductive history and retirement decisions among "younger" cohorts born 1940 or later. This is likely to indicate changes in the work-family nexus across time, which clearly deserves further attention.

References

Ahn, N., & Mira, P. (2002). A note on the changing relationship between fertility and female employment rates in developed countries. *Journal of Population Economics, 15*, 667–682.

Blanchet, D., Brugiavini, A., & Rainato, R. (2005). Pathways to retirement. In A. Börsch-Supan et al. (Eds.), *Health, ageing and retirement in Europe – First results from SHARE* (pp. 246–252). Mannheim: MEA.

Daly, M., & Rake, K. (2003). *Gender and the welfare state.* Cambridge: Polity Press.

Esping-Andersen, G. (1990). *The three worlds of welfare capitalism.* Princeton, NJ: Princeton University Press.

Hank, K. (2004). Effects of early life family events on women's late life labour market behaviour: An analysis of the relationship between childbearing and retirement in western Germany. *European Sociological Review, 20*, 189–198.

Henretta, J. C., O'Rand, A., & Chan, C. G. (1993). Joint role investments and synchronization of retirement: A sequential approach to couple's retirement timing. *Social Forces, 71*, 981–1000.

Martínez-Granado, M., & Mira, P. (2005). The number of living children. In A. Börsch-Supan et al. (Eds.), *Health, ageing and retirement in Europe – First results from SHARE* (pp. 48–52). Mannheim: MEA.

O'Rand, A. M., Henretta, J. C., & Krecker, M. L. (1992). Family pathways to retirement. In M. Szinovacz et al. (Eds.), *Families and retirement: Conceptual and methodological issues* (pp. 81–98). Beverly Hills, CA: Sage.

Pienta, A. M., Burr, J. A., & Mutchler, J. E. (1994). Women's labor force participation in later life: The effects of early work and family experiences. *Journal of Gerontology: Social Sciences, 49*, 231–239.

Chapter 15
Quality of Work, Health and Early Retirement: European Comparisons

Johannes Siegrist and Morten Wahrendorf

15.1 Relations of Quality of Work and Retirement

Extending labour marked participation of older people (aged 55+) is an important target of European social policy ("Lisbon Strategy"). In addition to reducing economic incentives of early retirement, investments into "good" work, in terms of a favourable psychosocial work environment, are proposed as promising measures towards this end. Distinct national social policies may enhance such efforts.

In this contribution, we investigate whether important aspects of a "good" quality of work (in terms of a favourable psychosocial work environment) experienced during a relevant stage of people's employment trajectories are associated with a reduced probability of early retirement. Preliminary findings based on SHARE data indicate that the intention to leave work and employment prematurely is strongly associated with a poor work environment, in particular with a health-adverse psychosocial work environment (Siegrist and Wahrendorf 2009; Siegrist et al. 2007). However, it is not known whether factual retirement decisions follow the same pattern. Because health is an important determinant of early retirement we are also interested in exploring associations of quality of work with health, at least, given the restrictions of our study design, with health status after labour market exit.

Given the fact that SHARE offers opportunities of studying country variations we additionally analyse associations of distinct indicators of national labour and welfare policies with quality of work as well as with early retirement. This analysis is based on evidence indicating differential effects of labour and welfare policies on quality of work and its association with health (Dragano et al. 2010). More specifically the following two complementary hypotheses are tested: First, we assume a relationship between the degree of active labour and social policy and aggregate measures of quality of work across the European countries under study. Second, we

J. Siegrist (✉) and M. Wahrendorf
Department of Medical Sociology, University of Duesseldorf, Universitätsstrasse 1, 40225 Duesseldorf, Germany
e-mail: siegrist@uni-duesseldorf.de; wahrendorf@uni-duesseldorf.de

assume a relationship between the degree of active labour and social policies and the extent of early retirement across countries.

How is quality of work defined and measured in this analysis? To measure quality of work theoretical models are needed that identify specific stressful job characteristics. Several such models were developed (Antoniou and Cooper 2005), but two models received special attention in occupational research, the demand-control-support model (Karasek et al. 1998) and the effort-reward imbalance model (Siegrist et al. 2004). The first model identifies stressful work by job task profiles characterised by high demand, low control (decision latitude) and low social support at work. The second model claims that an imbalance between high efforts spent and low reward received in turn (money, esteem, career opportunities, job security) adversely affects health. In SHARELIFE all core dimensions of these two work stress models were assessed using 12 Likert-scaled items from the original questionnaires (see Measurement).

Against this background, we provide preliminary answers to the following three questions:

1. Is poor quality of work experienced during a significant period of participants' employment trajectory associated with reduced health after labour market exit?
2. Is poor quality of work experienced during a significant period of participants' employment trajectory associated with a higher probability of early retirement?
3. Does quality of work vary according to specific indicators of national labour market and social policies? Does the same hold true for the probability of early retirement?

15.2 Measuring Quality of Work in SHARELIFE

In addition to retrospective data from the SHARELIFE project, we use data derived from the second wave of SHARE with information on respondents' health status. For the analyses, we included all people aged 50 or older who reported to be employed at least once during their life course. Furthermore, since we were interested in the influence of quality of work on health during retirement, we restricted the sample to people who already left the labour market in wave 2. Finally, respondents who had difficulties to respond to the retrospective questionnaire (4%) were not included either. This results in a sample of 6,619 men and 7,688 women (N = 14,307) from 13 European countries.

SHARELIFE contains an extensive module on work history collecting information on each job a respondent had during his or her working career (the mean number of jobs is 2.7; see also Chap. 11). In addition to general information (e.g. occupational status, working time), this module includes an assessment of the psychosocial work environment of the last main job of the working career (lasting longer than 5 years). As a result, quality of work during working life (assessed retrospectively) can be related to health and well-being during retirement. Furthermore, quality of work can be related to information on the participants' retirement behaviour. For the respective analyses, we created five binary indicators of poor quality of work, all based on 12 questionnaire items (4-point Likert scaled) taken from established work stress measures. Each indicator corresponds to a core dimension of existing work stress models (Karasek et al. 1998; Siegrist et al. 2004): physical demands (two items), psychosocial demands (three items), social support at work (three items), control at work (two items), and reward (two items). The respective dimensions were replicated in factor analyses. Here we calculate a simple sum-score for each dimension with higher scores indicating poorer quality of work (threshold: scoring in the upper tertile of the respective measure). The items are displayed in Table 15.1.

Our first research question points to the association of poor quality of work with health status after labour market exit. To this end, we used five different binary indicators of health, taken from the second wave of SHARE and widely used in recent publications (e.g. Avendano et al. 2009): Poor self-perceived health (less than good), scoring high on depressive symptoms (more than three symptoms on the EURO-D-scale), diagnosed chronic diseases (two or more), self-reported symptoms (two or more), and a measure of functional limitation (at least one ADL or IADL limitation). To study associations between quality of work and early retirement (our second research question), we created a binary indicator measuring whether or not respondents were employed at the age of 60 (for all respondents aged 60 or older).

We choose specific macro indicators related to labour market policies within the European countries under study, in particular measures of active labour market policies (ALMP). In general, six different categories of ALMP are distinguished (cf.: European Commission 2009), of which two are used in the context of our analyses: (a) measures related to training programs for the working population, and (b) measures related to rehabilitative services of a country. Training programs refer to programs aiming at increasing working skills, such as workplace training or further education. They improve the level of qualification and strengthen older people's position within the labour market. To represent this category, we use two indicators

Table 15.1 Measures of quality of work

Dimension	Item (strongly agree, agree, disagree, strongly disagree)
Physical demands	1. My job as [job title] during [time period] was physically demanding.
	2. My immediate work environment was uncomfortable (e.g. noise, heat, crowding).
Psychosocial demands	3. I was under constant time pressure due to heavy workload.
	4. My work was emotionally demanding.
	5. I was exposed to recurrent conflicts and disturbances.
Control	6. I had very little freedom to decide how to do my work.
	7. I had an opportunity to develop new skills.
Reward	8. I received the recognition I deserved for my work.
	9. Considering all my efforts and achievements, my salary was adequate.
Social support at work	10. I received adequate support in difficult situations.
	11. There was a good atmosphere between me and my colleagues.
	12. In general, employees were treated with fairness.

in our analyses, one indicator referring to the factual participation in such activities, and one indicator referring to the extent of a country's labour market expenditures invested into training programs. Specifically, the first indicator is measured as percentage of persons aged 25–64 who stated that they received education or training in the last month. The extent of expenditures is measured as percentage of GDP. The second category of ALMP concerns rehabilitative services in a country, and more specifically supported employment and rehabilitation services for people with limited working capacity. Such measures are thought to increase rates of return to work of people with chronic illness and to reduce time intervals from treatment to re-uptake of work. Our proposed indicator is the amount of a country's expenditures in such programs, expressed as percentage of GDP. For each macro-variable, we collected information available from 1985 to 2005 from the OECD database, and we computed a respective country mean score for each indicator.

Additional variables are gender, age (divided into age categories), and occupational status (based on ISCO-codes) of the main job of the working career. Respective categories are "legislators and professionals", "associated professionals and clerks", "skilled workers", and "elementary occupations".

We performed two sets of analyses. First, we present bivariate and multivariate associations between poor quality of work on the one hand and health and early retirement on the other hand. All multivariate analyses are based on logistic multilevel models for binary outcome variables, with individuals (level 1) nested within countries (level 2). This allows for an accurate adjustment for country affiliation. In a second set of analyses, we investigate associations between the three macro indicators on the one hand and the two main measures of quality of work (low control, low reward) and early retirement on the other hand. Weights were considered within the analyses.

15.3 Effects of Quality of Work on Health and Early Retirement

Is the experience of poor quality of work during a prolonged period of one's working associated with poor health after labour market exit? Table 15.2 gives an initial answer to this question – our first research question. Results indicate that people who experienced poor quality of work in their main job are more likely to report reduced health. This holds true for all five indicators of poor health. As exemplified in Fig. 15.1 by the health indicator of depressive symptoms, this association holds true for all single countries under study, using low control and low reward at work as two main indicators of poor quality of work. The associations reported in Table 15.2 remain significant in multilevel models, where age, gender, occupational status, and country affiliation are considered as potential confounders (see Table 15.2). These findings suggest that the experience of an adverse psychosocial

Table 15.2 Quality of work and health after labour market exit and early retirement (in %)

		Poor health					Empl. at age 60
		Poor SRH	High depr. symptoms	2+chronic diseases	2+symptoms	Functional limitations	
High phys. demands	Yes	56.2	35.1	61.4	57.4	29.2	19.9
	No	41.4	25.5	49.5	45.4	18.9	25.4
High psych. demands	Yes	48.6	25.9	55.5	52.5	23.6	21.3
	No	43.2	32.2	50.9	46.3	20.2	25.1
Low work control	Yes	55.9	24.5	50.5	57.2	30.0	17.4
	No	41.5	39.1	57.8	45.5	18.7	26.0
Low reward	Yes	51.9	24.5	57.7	56.5	26.7	21.2
	No	42.7	38.7	50.6	45.7	19.6	25.1
Low social support	Yes	50.9	36.8	49.2	54.1	25.6	20.6
	No	40.7	23.3	56.1	44.4	18.1	22.9

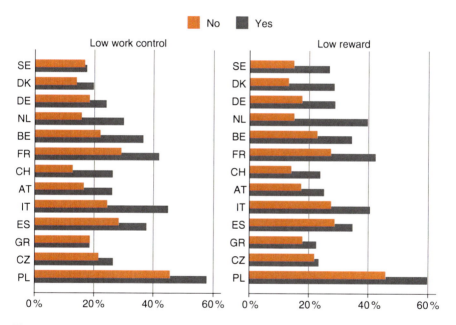

Fig. 15.1 Quality of work and depressive symptoms after labour market exit

environment is related to poor health during retirement – even after taking into account workers' occupational status and country affiliation.

To investigate our second research question, – the association between quality of work and probability of early retirement – we again consider Table 15.1 where the proportion of people still employed at the age of 60 is presented according to the five indicators of poor quality of work. As can be seen, this proportion of retired people who indicated to have been still employed by the age of 60 is always higher in the group reporting good quality of work as compared to the group reporting poor quality of work. However, once the multivariate model adjusts for the effects of occupational status and country affiliation these differences are no longer statistically significant, with the exception of low control at work. This observation may indicate that occupational status accounts for some part of the association between poor quality of work and early retirement. Similarly, the country seems to be an important confounder, affecting both the level of quality of work and the probability of being still employed at the age of 60. This latter aspect is explored more rigorously in the following section.

How are the three macro indicators related to the two main indicators of poor quality of work (low control and low reward), and to early retirement? Answers are given in Figs. 15.2 and 15.3. First, we observe a pronounced association between a country's amount of activities related to lifelong learning and its aggregate measure of quality of work (on the left hand side of Fig. 15.2): Higher participation rates in lifelong learning go along with better mean quality of work (higher amount of control and reward at work). However, when comparing this ALMP indicator with the second indicator, the proportion of expenditures in such activities as part of the GDP, respective associations with quality of work are less pronounced (see the

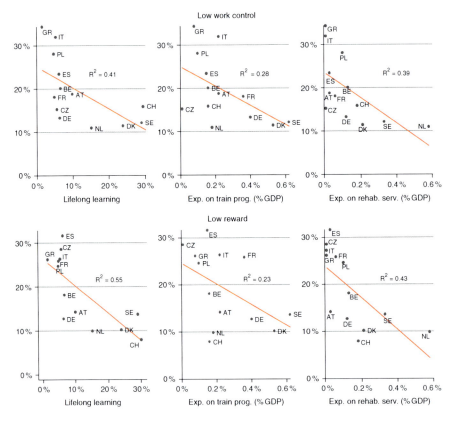

Fig. 15.2 Macro indicators and poor quality of work

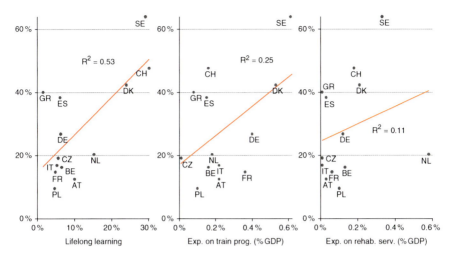

Fig. 15.3 Macro indicators and still employed at age 60

Table 15.3 Associations between quality of work and health and early retirement: results of multilevel logistic regression models (odds ratios and level of significance)

		Poor health					Empl. at age 60
		Poor SRH	High depr. symptoms	2+chronic diseases	2+symptoms	Functional limitations	
High phys. demands	Yes	1.40***	1.29***	1.43***	1.47***	1.46***	0.94
	No	–					
High psych. demands	Yes	1.32***	1.40***	1.32***	1.49***	1.43***	0.95
	No	–					
Low work control	Yes	1.33***	1.36***	1.10*	1.28***	1.31***	0.77***
	No	–					
Low reward	Yes	1.33***	1.55***	1.18***	1.42***	1.31***	1.03
	No	–					
Low social support	Yes	1.45***	1.64***	1.29***	1.46***	1.47***	1.03
	No	–					

All odds ratios are based on logistic multilevel models with individuals nested in countries, and are adjusted for age categories, gender, and occupational status in main job
Note: $*p < 0.05$; $**p < 0.01$; $***p < 0.001$

middle of Fig. 15.2). This finding may indicate that the first variable is better suited to capture a respective macro-level effect on quality of work (Table 15.3).

Second, with regard to expenditures in rehabilitative services, associations with quality of work are again observed in the expected direction: higher investments at country level go along with better mean quality of work (see right hand side of Fig. 15.2).

Third, in Fig. 15.3 we analyse respective associations of the three macro-indicators with the probability of staying at work beyond age 60. Again, the strongest associations are found in case of rates of participation in lifelong learning, where continued employment at older age is more prevalent among people working in countries with high rates of participation, such as Sweden, Denmark or the Netherlands.

15.4 Summary

In this chapter, retrospective data from SHARELIFE were used to measure poor quality of work in working life and to study its association with five indicators of poor health after labour market exit (question 1), and with continued employment in late mid-life (question 2). Furthermore, we studied whether poor quality of work and continued employment vary according to specific macro indicators of labour market policies (question 3).

Our main results are as follows:

- First, we found strong evidence that people who experienced poor working conditions during a significant period of their employment trajectory are more likely to report poor health during retirement. Associations were consistent across different health indicators and were observed for all five indicators of poor quality of work (high physical demands, high psychosocial demands, low control, low reward, and low social support at work). Associations remain significant after considering occupational status and country-affiliation in multivariate analyses. Apparently, poor quality of work remains associated with people's health status after retirement, independent of occupational status and country affiliation.
- Second, continued employment at older age (60 or older) was found to be more prevalent among people who experienced "good" quality of work, in particular high control at work. However, these associations weakened considerably when occupational status and country affiliation were considered in multivariate analyses. This suggests that occupational status and country affiliation both affect the level of quality of work and the probability of being still employed at the age of 60.
- Third, quality of work was generally higher in countries with a pronounced active labour market policy. This association was most pronounced in case of high participation rates in training programs for adults (lifelong learning). Similarly, continued employment into old age was more prevalent in countries with high expenditures in rehabilitation services.

In conclusion, these results show that an active labour policy for older workers and the investment into continued education during working life (life long learning) have beneficial effects on working conditions, in terms a favourable psychosocial work environment. Given the strong associations of good quality of work on mental and physical health, long-term effects on employees' health are considerable. Therefore, promoting quality of work by strengthening these more distant determinants may have beneficial medium- and long-term effects on the workability of an ageing workforce in Europe.

References

Antoniou, A. S., & Cooper, C. (2005). *Research companion to organizational health psychology*. Chelterham: Edward Elgar Publishers.

Avendano, M., Jürges, H., & Mackenbach, J. P. (2009). Educational level and changes in health across Europe: Longitudinal results from SHARE. *Journal of European Social Policy, 19*, 301–316.

Dragano, N., Siegrist, J., & Wahrendorf, M. (2010). Welfare regimes, labour policies and workers' health: A comparative study with 9917 older employees from 12 European countries. *JECH*, in press.

European Communities. (2009). *Labour market policy – Expenditures and participants*. Luxembourg: European Union.

Karasek, R., Brisson, C., Kawakami, N., Houtman, I., Bongers, P., & Amick, B. (1998). The Job Content Questionnaire (JCQ): An instrument for internationally comparative assessments of psychosocial job characteristics. *Journal of Occupational Health Psychology, 3*, 322–355.

Siegrist, J., & Wahrendorf, M. (2009). Quality of work, health, and retirement. *The Lancet, 374*, 1872–1873.

Siegrist, J., Wahrendorf, M., von dem Knesebeck, O., Jürges, H., & Börsch-Supan, A. (2007). Quality of work, well-being, and intended early retirement of older employees – Baseline results from the SHARE Study. *The European Journal of Public Health, 17*, 62–68.

Siegrist, J., Starke, D., Chandola, T., Godin, I., Marmot, M., Niedhammer, I., et al. (2004). The measurement of effort-reward imbalance at work: European comparisons. *Social Science & Medicine, 58*, 1483–1499.

Chapter 16
Working Conditions in Mid-life and Participation in Voluntary Work After Labour Market Exit

Morten Wahrendorf and Johannes Siegrist

16.1 Social Position and Participation Across European Countries

Promoting participation in productive activities after labour market exit is an important challenge for European policies. Not only society as a whole might profit from an increased investment, but also older people themselves, since participation in a productive activity, such as voluntary work, was shown to improve health and well-being in older ages (Bath and Deeg 2005) – a finding that was also found in the two first waves of SHARE (Siegrist and Wahrendorf 2009a). These results suggest that being engaged in a productive activity after labour market exit helps to cope with the ageing process because valued earlier activities are replaced by new ones, providing opportunities of positive self-experience which in turn strengthens well-being and health. Previous findings, though, show that participation varies considerably according to social position and between different countries (see also Hank 2010).

Higher participation rates were particularly observed among people with higher education and higher income. Whereas the descriptive evidence of this social gradient of participation is convincing, the explanations given so far are limited and still need to be explored. In particular, the relative contribution of earlier stages in the life course in explaining these variations, such as working conditions in mid-life, remains an open challenge. But why – or how – should former working conditions exactly be related to participation in productive activities after labour market exit?

One reason might be that the motivation of getting active is higher, given that positive work-related experience occurred. In other words, people in low social position might experience poor working conditions in terms of exposure to psychosocial stress at work. As a result, the intention to retire is higher (Siegrist and

---

M. Wahrendorf (✉) and J. Siegrist
Department of Medical Sociology, University of Duesseldorf, Universitätsstrasse 1, 40225 Duesseldorf, Germany
e-mail: wahrendorf@uni-duesseldorf.de; siegrist@uni-duesseldorf.de

Wahrendorf 2009b), and leaving the labour market is probably not experienced as such a remarkable "role loss", which needs to be compensated, but rather as a relief from the obligations of employment (Westerlund et al. 2009). As a consequence, people in low social position are probably less willing to engage in such an activity after labour market exit. Another reason might be that "good" working conditions in midlife contribute to increased health at older age, which in turn favours the participation in productive activities after labour market exit. Along these lines, several longitudinal investigations demonstrate that former working conditions exert long-lasting effects on health and well-being (Blane 2006). But long lasting influences on participation in productive activities still need to be explored.

In addition to variations according to social position, former SHARE-findings also show that participation rates differ between countries, in particular in case of volunteering (Hank 2010; Siegrist and Wahrendorf 2009a). While rates of volunteering were found to be high in the Northern countries together with the Netherlands, Belgium and France, rates were rather low in Italy, Spain and Greece and the two Eastern countries. So far, these findings are generally discussed in the frame of tailored policy programs which may encourage participation in productive activity (Hank 2010; Salomon and Sokolowski 2003). But which policy programs are these exactly? Can this probably also be related to country-specific labour market policies that increase working conditions in mid-life (e.g. rehabilitative services). Recent findings based on the SHARE study show clear country-variations of quality of work according to such factors (Dragano et al. 2010). But can these factors also be related to participation after labour market exit?

Taken together, the complex associations between policy measures, working conditions in mid-life and participation in productive activities remain as an open challenge and have not been explored so far. One reason is that former investigations are mainly based on cross-sectional findings, with no available information from former stage in the life course. With its broad set of retrospective life history information from respondents in 13 Continental European countries, SHARELIFE offers unique opportunities to relate former life stages with participation after labour market exit, and to give initial answers to the addressed questions above. More specifically, three questions will be explored.

1. Are working conditions in mid-life associated with participation in productive activities after labour market exit?
2. If so, to what degree can this association be explained by better health after labour market exit?
3. Which macro factors are related to higher participation rate and might help to increase participation in productive activities in older ages?

To study these questions, we focus on volunteering as an important type of productive activity, and we analyse working conditions, in terms of different aspects of respondents' work history (see Measurement section), including the exposure to psychosocial stress at work during the working career – all information taken from the retrospective data collection in SHARELIFE.

16.2 Measuring Working Conditions in Mid-life in Relation to the Welfare State

For our analyses we combined the information from the SHARELIFE survey, with information from the second wave of SHARE, where participation in volunteering was assessed. Since we were interested in participation in voluntary work after labour market exit, we restricted the sample to people aged 50 or older who already left the labour market at wave 2. Moreover, if the interviewer reported any difficulties of the respondent to answer the questions in the retrospective questionnaire, respondents were excluded (4% of the cases). This restriction results in a sample of 14,150 respondents. Weights were considered within the analyses.

SHARELIFE includes an extensive module on work history. This module allows us to reproduce the respondents' principal occupational situation from the age of 15 onwards, by collecting information on each job together with details on periods where the respondent was not employed (if the respective period lasted 6 months or longer). Information on jobs includes a measure of occupational status (based on ISCO codes), information on working time (full-time or part-time), and information on the psychosocial work environment (for the last main job of the working career). Information on existing gaps includes a description of the situation (e.g. unemployed, sick or disabled, domestic work, etc.). On this basis, we created a variable describing the respective employment situation for each age between 15 and 65 (or age at the SHARELIFE interview if respondent younger than 65) using seven different categories. The categories and their prevalence by age are displayed in Fig. 16.1 separately for men and women for the total SHARELIFE sample.

For the following analyses, six variables were created to describe specific characteristics of respondents' work history based on the collected information: (1) a binary indicator describing whether the respondent experienced an episode of unemployment in working life, (2) years since labour market exit, (3) a variable indicating whether an episode of being sick and disabled occurred during working life, (4) the average number of job changes in the working life. Moreover, we created a variable measuring (5) the occupational status of the main job of the working career. The categories are "Legislators and professionals", "Associated professionals and clerks", "skilled workers", and "elementary occupations". And lastly, to measure adverse psychosocial working conditions, we created (6) five binary indicators measuring core dimensions of work stress – again for the main job of the working career. Those indicators are based on 12 questionnaire items (four-point Likert scaled) referring to existing questionnaires (Karasek et al. 1998; Siegrist et al. 2004). The assessed dimensions are physical demands (two items), psychosocial demands (three items), social support at work (three items), control at work (two items), and reward (two items). For the analyses, we calculated a simple sum-score for each dimension with higher scores indicating higher work stress and created a binary indicator, where participants scoring in the upper tertiles of the respective measure were considered experiencing poor quality of work.

Rather than using existing welfare state typologies for our analyses, we choose specific macro indicators of welfare state interventions of a country, both related to

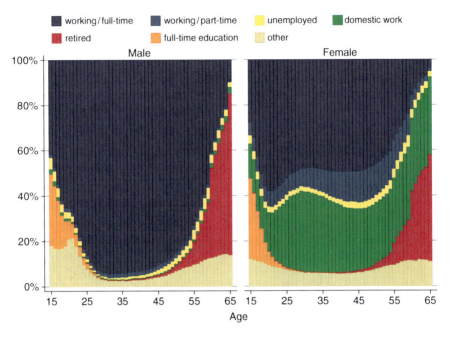

Fig. 16.1 Employment situation by age

the labour market policy of a country: (1) extent of lifelong learning, and (2) amount of expenditure in rehabilitative care. Life long learning is a key concept to promote decent employment at all ages. Especially the older workforce profits from continuous education as it improves their level of qualification and therefore their position in the labour market. The variable refers to persons aged 25–64 who stated that they received education or training in the 4 weeks preceding the EU Labour Force Survey. Information on expenditure in rehabilitation service is taken from the Eurostat database on labour market policy. It refers to the labour market policy expenditures that are invested in supported employment and rehabilitation services, measured as percentage of GDP. For both macro-variables, we used the mean value based on the available information since 1985. Both aspects are thought to be related to "good" working conditions – life long learning, since it improves the level of qualification and increases personal control at work and opportunities in the labour market, and rehabilitative care, since it influences the probability and time interval of returning to work by increasing the health status during working life.

As mentioned above, participation in voluntary work was used as our main outcome variable, which was measured in wave 2 of the SHARE project. More specifically, respondents were asked whether or not they were involved during the last 4 weeks in "voluntary or charity work".

Additional variables are gender, age (divided into age categories), education (low, medium, high) as well as two health indicators taken from wave 2: one binary indicator for functional limitation (either ADL or IADL limitations) and poor self-perceived health (less than good on a five–likert scale ranging from excellent to poor).

16.3 Associations of Working Conditions and Volunteering After Labour Market Exit

How are working conditions related to participation in volunteering? Table 16.1 gives an initial answer to this question, by showing the participation rates according to all covariates under study. But before turning to working conditions, we first observe that participation varies according to gender (more men), age (higher rates between age 60 and 80), and education (higher rates among higher educated). Furthermore, both health indicators are associated with increased participation rates. With regard to working conditions the results suggest that participation

Table 16.1 Participation in voluntary work according to covariates in % (weighted)

Gender	Male	13.6
	Female	11.4
Age group (wave 2)	50–59 years	10.7
	60–69 years	14.7
	70–79 years	12.7
	80 years or more	6.0
Education	Low	7.5
	Medium	13.8
	High	24.7
Functional limitations	Yes	6.3
	No	14.0
Poor self-rated health	Yes	7.8
	No	16.1
Occupational status	Legislators and professionals	25.6
	Ass. professionals and clerks	15.0
	Skilled workers	11.0
	Elementary occupations	7.0
Episode of unemployment	Yes	9.8
	No	12.7
Episode of sick and disabled	Yes	9.6
	No	12.4
Job changes	None	9.3
	1–2	12.8
	3 or more	15.4
Years since last job	1–5 years	15.5
	5–15 years	13.6
	16 or more	9.9
High physical demands	Yes	5.7
	No	14.0
High psychosocial demands	Yes	11.8
	No	12.4
Low work control	Yes	7.1
	No	13.6
Low reward	Yes	7.5
	No	13.5
Low social support	Yes	9.0
	No	14.1
Total		12.4

rates are higher among people who had a higher occupational status in working life, who experienced frequent job changes in their, and among those who experienced no episode of unemployment or of being sick and disabled.

When turning to the indicators of psychosocial work conditions, we observe for all five indicators that people who experienced poor working conditions in their main job are less likely to participate in voluntary work once they left the labour market. This holds particularly true in case of high physical demands, low work control, low reward at work and in case of low social support at work. Results of Table 16.1 are consistent at the country level – as exemplified for social position (Fig. 16.2) and for low control and low reward (Fig. 16.3).

But to what extent can these associations be explained by increased health in older ages? To answer this question (our second research question) we additionally present results of multivariate analyses in Table 16.2, where the effect is estimated for each working condition – before (model 1) and after adjustment for the two health indicators under study (model 2). When turning to model 1 (first column), we observe that findings of the descriptive analyses remain stable. Again, people with a higher occupational status in their working career are more likely to participate in volunteering during retirement, as well as people who experienced no period of unemployment, or with frequent job changes. Furthermore, four of the five work stress indicators are found to be significantly associated to volunteering during retirement, namely high physical demands, low work control, low reward at work and low social support at work. Importantly all these reported associations are only modestly reduced when adjusting for health after labour market exit in model 2, and

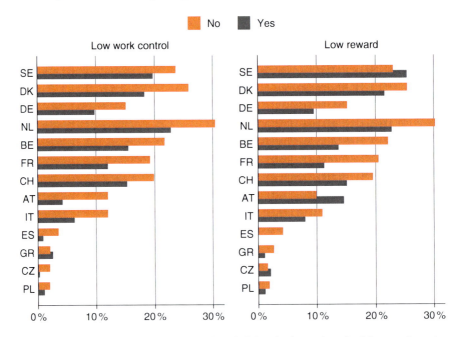

Fig. 16.2 Psychosocial working conditions in main job and volunteering after labour market exit

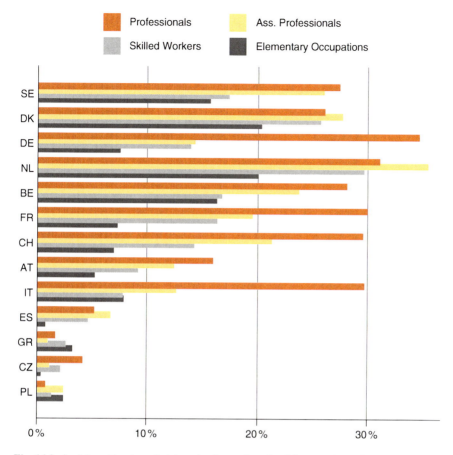

Fig. 16.3 Social position in main job and volunteering after labour market exit

these associations remain significant. This result suggests that the experience of poor working conditions in midlife reduces the probability of volunteering during retirement independent of participants' health status.

Next, we explore how the two macro indicators are related to country-rates of participation. Results are displayed in Fig. 16.4. We observe that participation rates are generally lower in countries with lower rates of life long learning and in countries that invest less in occupational rehabilitative services.

16.4 Summary

With this contribution we set out to study the long lasting influences of working conditions in mid-life on participation in voluntary work after labour market exit using information on respondents' work history collected in the SHARELIFE

Table 16.2 Associations between working conditions and participation in voluntary work: results of multilevel logistic regression models (odds ratios and significant level)

		Model 1	Model 2
Occupational status	Legislators and professionals	1.54***	1.47***
	Ass. professionals and clerks	1.43***	1.36***
	Skilled workers	1.28**	1.25**
	Elementary occupations		
Episode of unemployment	Yes	0.77**	0.78*
	No		
Episode of sick and disabled	Yes	0.78	0.89
	No		
Job changes	None		
	1–2	1.14*	1.15*
	3 or more	1.31***	1.33***
Years since last job	1–5 years		
	5–15 years	0.97	1.00
	16 or more	0.81*	0.86
High physical demands	Yes	0.72***	0.76**
	No		
High psychosocial demands	Yes	1.00	1.04
	No		
Low work control	Yes	0.79**	0.83*
	No		
Low reward	Yes	0.72***	0.75***
	No		
Low social support	Yes	0.79**	0.84*
	No		

Model 1: adjusted for age categories, gender, and educational attainment
Model 2: Model 1 plus self-perceived health and functional limitations
Note: *$p < 0.05$; **$p < 0.01$; ***$p < 0.001$

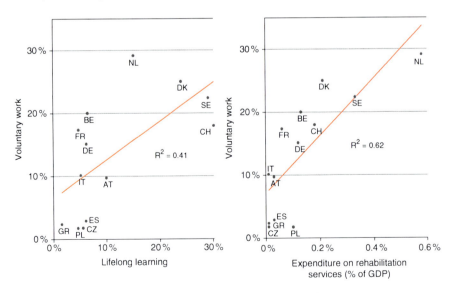

Fig. 16.4 Macro indicators and volunteering after labour market exit

interview to identify specific working conditions in mid-life. By doing so, we were particularly interested in studying the relative contribution of an adverse psychosocial work environment on the probability of participation. Moreover, we studied the question to what extent this association can be explained by decreased health status during retirement. Finally, we explored the relation between specific macro indicators which are thought to affect active engagement of retired people (by improving work and employment in midlife) and likelihood of participating in volunteering. Results can be summarized as follows:

1. Our findings emphasize that people who experienced poor working conditions in midlife are less likely to engage in volunteering after labour market exit. This holds true for people who experienced an episode of unemployment, who were holding a low status job, and who had few job changes. Moreover, poor psychosocial working conditions were associated with low participation rates, specifically work and employment conditions defined by high physical demands, low control, low reward, and low social support.
2. These reported associations remain significant after adjusting for health status (functional limitations and self-perceived health). Apparently, working conditions seem to have long-lasting effects on the probability of participating in productive activities after labour market exit – effects that are not substantially reduced by people's health status.
3. The extent of volunteering in early old age is also influenced by a nation's investments into macro-structural policy measures that aim at improving quality of work and employment. Our results demonstrate this effect for two such indicators, the extent of lifelong learning, and the amount of resources spent in rehabilitation services. In either case, country's overall rates of participation in volunteering were clearly higher compared to countries with fewer investments.

Our results show that SHARELIFE offers promising opportunities to study life course influences on participation in productive activities after labour market exit, focussing on working conditions over the life course. Previous investments in voluntary work during working life (beside work) might be another important predictor (Erlinghagen 2010). Yet, since we focussed on respondents' principal occupational situation over the life course, this information was not included. Despite this possible shortcoming, the findings suggest that promoting working conditions in midlife might not only increase health and well-being, but also encourage participation in productive activities after labour market exit.

References

Bath, P. A., & Deeg, D. (2005). Social engagement and health outcomes among older people: Introduction to a special section. *European Journal of Ageing, 2*, 24–30.
Blane, D. (2006). The life course, the social gradient, and health. In M. Marmot & R. Wilkinson (Eds.), *Social determinants of health* (pp. 54–77). Oxford: Oxford University Press.

Dragano, N., Siegrist, J., & M. Wahrendorf, (2010). Welfare regimes, labour policies and workers' health: A comparative study with 9917 older employees from 12 European countries. *JECH*, in press.

Erlinghagen, M. (2010). Volunteering after retirement. Evidence from German panel data. *European Societies, 12*(5), 603–625.

Hank, K. (2010). Societal determinants of productive aging: A multilevel analysis across 11 European countries. *European Sociological Review*, in press.

Karasek, R., Brisson, C., Kawakami, N., Houtman, I., Bongers, P., & Amick, B. (1998). The Job Content Questionnaire (JCQ): An instrument for internationally comparative assessments of psychosocial job characteristics. *Journal of Occupational Health Psychology, 3*, 322–355.

Salomon, L. M., & Sokolowski, S. W. (2003). Institutional roots of volunteering: Towards a macro-structural theory of individual voluntary action. In L. Halman & P. Dekker (Eds.), *The values of volunteering* (pp. 71–90). New York: Kluwer Academic.

Siegrist, J., Starke, D., Chandola, T., Godin, I., Marmot, M., & Niedhammer, I. (2004). The measurement of effort-reward imbalance at work: European comparisons. *Social Science & Medicine, 58*, 1483–1499.

Siegrist, J., & Wahrendorf, M. (2009a). Participation in socially productive activities and quality of life in early old age: Findings from SHARE. *Journal of European Social Policy, 19*, 317–326.

Siegrist, J., & Wahrendorf, M. (2009b). Quality of work, health, and retirement. *The Lancet, 374*, 1872–1873.

Westerlund, H., Kivimäki, M., Singh-Manoux, A., Melchior, M., Ferrie, J., & Pentti, J. (2009). Self-rated health before and after retirement in France (GAZEL): A cohort study. *The Lancet, 374*, 1889–1896.

Part III
Health and Health Care

Chapter 17
Scar or Blemish? Investigating the Long-Term Impact of Involuntary Job Loss on Health

Mathis Schröder

17.1 Job Loss: A Risk for Individual Wellbeing

Job loss is one of the most undesirable events for an individual, as it may affect a person's wellbeing negatively on a multitude of dimensions. There may be direct and indirect effects of job loss on wellbeing, and these effects may be temporary or permanent, depending on the outcome of interest. Ruhm (1991) and Jacobson et al. (1993) for example show that wage earnings – not surprisingly – decline dramatically in the event of a job loss. However, this effect is permanent in the sense that a comparable group without a job loss follows a much higher wage path in the years following the job loss. Effects on family life have also been shown: Charles and Melvin (2004) find a significant increase in divorce probability for laid-off individuals. Bono et al. (2008) show that fertility decisions are changed due to a job loss, resulting in delayed or forgone child birth. Various papers find a negative effect of job loss on health. Clearly, health is an example where an indirect effect can be imagined – through lower wages, investments in health are declining, and then overall health is affected in addition to a possible direct effect (through depressive symptoms caused by the unemployment situation, for example). Sullivan and von Wachter (2009) find an increase in mortality, Strully (2009) shows that there are effects on subjective health measures as well. Others, for example Salm (2009) or Browning et al. (2006), argue that health effects are rather spurious and not necessarily caused by the job loss. Long-term effects on health have not yet been investigated, mainly due to lack of suitable data.

One common problem in the literature on involuntary job loss is the direction of causality: if people are less productive, they will lose their job and then are likely to earn less in the future. A similar argument holds for deteriorating health, which might lead to a loss of the job in the first place. A solution to this issue is to find reasons for the job loss, which are not linked to the outcome variable. The literature cited above uses the common approach of "displaced" workers to (arguably) solve

---

M. Schröder
German Institute for Economic Research (DIW) Berlin, Mohrenstr. 58, 10117 Berlin, Germany
e-mail: mschroder@diw.de

this problem. A displaced worker is a worker who has lost his job due to a plant closure, where the plant is sufficiently large for the individual worker not to matter much in the closure (Ruhm 1991; Jacobson et al. 1993; Sullivan and von Wachter 2009). In this case, the individual's influence is not enough to cause the closure, but the closure affects all those losing their job in the same way. Even though this approach may not be perfect, it is the closest way to disentangle the individual reasons for job loss from a common and thus exogenous one.

The effects of an exogenous involuntary job loss, however, do not only depend on the loss of the job itself, but to a potentially large extent also on the institutional settings in the background. In a system with mandatory complete health insurance coverage, one would expect the effects of job loss on health to be less severe than in a system which only provides health insurance coverage through the employer. Institutional settings may also shorten the duration in unemployment after job loss, which could be positive or negative on wages – a higher match quality may take more time to be found but lead to higher wages, but a long lasting unemployment support may lead to larger human capital depreciation. Where a family is supported – financially or otherwise – independent of employment of its members, a spell of unemployment may have less influence on fertility or marriage decisions.

This paper uses the SHARELIFE data to investigate the long term effects of job loss on health, exemplified through three health outcomes, to test whether health effects last or are only temporary. The effects of job loss due to displacement are compared to those which are related to lay-off or firing to test whether there are notable differences. Finally, a country's institutional settings related to the labour market are considered in their influences on individuals' long term health. SHARE-LIFE is especially useful for this project, as it allows for the first time to relate events in a person's life course (here: involuntary job loss) to long term outcomes (here: health). A combination of SHARELIFE data with SHARE data from waves 1 and 2 is used for a full set of control variables.

The next section provides a review of the literature. Section 17.3 defines the sample, explains the measurements of the outcome variables and introduces the institutional variables that are used in this analysis. Section 17.4 then shows the analyses, whereas Sect. 17.5 briefly summarizes.

17.2 Literature Findings on Job Loss and Wellbeing

Ruhm (1991) uses five waves of the Panel Study of Income Dynamics (PSID) to investigate the effect of job loss on unemployment probability and wages. Applying a definition of mass layoffs and plant closures for displacement, he finds that while the unemployment probability in the following five years is not affected by displacement, wages are permanently lower by 10–13% compared to those workers not displaced. In a similar study, Jacobson et al. (1993) use administrative data from the US federal state of Pennsylvania to elicit the effects of mass-layoffs on wages of

17 Scar or Blemish? Investigating the Long-Term Impact of Involuntary   193

high-tenured workers. They find losses in earnings of about 25% compared to pre-displacement earnings even six years after the job was lost.

Sullivan and von Wachter (2009) employ the same Pennsylvania dataset to investigate how the health of these workers has suffered from displacement. In linking their sample to administrative death records they are able to test whether the mortality of displaced workers is increased after job loss. Indeed they find that mortality rates for the displaced workers are 50–100% higher than for the non-displaced. Strully (2009) uses three waves of the PSID (1999, 2001, 2003) when looking at the relationship of health and job loss. She distinguishes between several different job loss categories, allowing for *no-fault, fired or laid off, voluntary*, or *other* types of job loss. In a first step, Strully then relates health prior to the job loss to the job loss category and finds that while those *fired or laid off* are in worse health prior to the job loss, a *no-fault* job loss is not associated with bad prior health, which is taken as support for the exogeneity of plant closure and health. She reports that while the short-term effects for fired individuals are stronger on self-rated health, they are still substantial and significant for those with a *no-fault* job loss, suggesting that job loss is associated with worse health.

There are also some findings in the literature that question the effects of job loss on health. Salm (2009) looks at the health of individuals experiencing a job loss in the US with the Health and Retirement Study (HRS). He is able to look at objective as well as subjective measures of health by different reasons of job loss. He does not find any effect of plant closure or of being laid off on health, only – not surprisingly – when people leave for health reasons, their health deteriorates in the subsequent periods. The sample of investigation, however, consists only of older individuals and therefore the number of people "under treatment" of displacement is rather small. Browning et al. (2006) investigate how job loss is associated with medical stress indicators. They use a 10% random sample of the Danish male population and link these records to business and hospital records. They report that displacement does not cause hospitalization for stress-related illnesses for various definitions of displacement. They speculate that this finding may be related to the generous welfare scheme in Denmark compared to the United States. In another study of European administrative data, Eliason and Storrie (2009) look at the Swedish case and observe a higher mortality of those who are displaced in Sweden – thus contradicting the effects Browning et al. find in a country that at least from the outset is very similar to Denmark.

This chapter explores the possibilities of advancing the literature on two fronts: first, we analyze true long-term health effects after job loss. The SHARELIFE sample allows looking at individuals whose job changes have happened up to 50 years prior to the current health measures. This gives a new perspective on the consequences of involuntary job loss on health. In addition, for the first time we are able to compare the effects of displacement across a number of European countries, which allows investigating the influence of the institutional settings in these countries regarding the treatment of the unemployed. While the comparison of the previous studies across countries is difficult because the measures are different both for displacement and health, the ex-ante harmonization in the SHARELIFE countries guarantees that there are no such issues: the questions and health measures are identical across all participating countries.

17.3 Job Loss and Health in the SHARELIFE Sample

This paper uses SHARELIFE data, but augments it with the data from waves 1 and 2 from SHARE. Similar to the literature cited above, there are several restrictions to the sample, which are necessary to make sure that the effects found are not due to sample composition. The first selection has to be on those individuals who are at risk of losing their jobs due to displacement, which, obviously, rules those out who have never been employed. Furthermore, since civil servants and military personnel usually do not face the risk of a business closure, these are excluded from the analysis. Although self-employed face the risk of displacement, they are excluded in the analyses, because a business closure can be caused by their behaviour, and thus the causation of displacement on outcome is not (as) clear. In addition, as the agricultural sector has usually a lot more day-to-day workers and less stable employment, all individuals in this sector are left out.

There are a number of other restrictions to ensure that the analysis is sound. Only individuals between 50 and 90 years of age at the point of the third interview are considered, although the findings are robust against changes in this variable. To not be biased by individuals who change their jobs after training, the minimum age at job change is 20, whereas no person displaced after 60 is still in the data to avoid confounding effects with retirement decisions. To avoid effects of the depression and war era, only job changes after 1950 are used, and to avoid effects correlating with SHARE wave 1 and 2, only displacement up to 2004 are considered. Individuals of the former German Democratic Republic, of the Czech Republic and of Poland are not considered in the analysis, as they did not have comparable conditions before and after the fall of the communist countries. As events are sometimes a long time ago for these respondents, individuals with low cognitive ability (as measured in wave 2 of SHARE) are excluded.

There are two sets of individual explanatory variables: the first is to control for differences prior to job change. These include childhood variables, specifically health as a child, access to health care as a child, school performance, amenities at home, people per room, whether the home was private or not, and – as a measure for socio economic status of the parents – whether there were enough books to fill one shelf in the house at age 10. Variables prior to job change also include the years of education, occupation and industry indicators, and some parental characteristics. Job and industry characteristics are taken from the first job a person had to avoid confounding effects that may have come after job loss. The second set provides contemporary information in order to correct for differences at the point of the interview which are not influenced by displacement. These are age and gender, and – unless noted otherwise – country fixed effects.

The central explanatory variable is the one defining involuntary job loss and the comparison category. The variable RE031, which is asked for all job spells a person reports, provides an intuitive definition of three types of job changers (one voluntary, two involuntary):

1. Those who always changed their job on their own account and thus always voluntary
2. Those who were at least once laid off and never experienced a business closure
3. Those who have at least once lost their job because of a plant closure

In the analysis, the comparisons always use the voluntary changers as a benchmark for the other two groups, with the expectation that group (1) fares better than group (2) and (3), while the effects for group (2) are expected to be larger than for group (3) because of the before mentioned issues of causality.

The analyses are restricted to the following outcome variables – self-perceived health, defined as being in poor or fair health, having at least one chronic disease and having at least one depressive symptom as measured via the EURO-D scale in wave 2 of SHARE. This spread provides a nice overview of subjective, objective and mental health, thus covering a wide range of individual health measures. The effects of job loss depend on the institutional settings and on how well the individual is caught in the safety net the institutions provide. For this paper, the concentration is on the unemployment benefit structure that each country had in 2007, where four dimensions are considered: the replacement rate, the duration of payment, the qualifying conditions and the employee contributions. Assuming that the changes over time are not dramatic, using the current setting as a proxy adjusts reasonably well for the institutional background at the time of the job loss.

Figure 17.1 shows the differences in self perceived health comparing those losing their job involuntarily and those who change voluntarily, split by country.

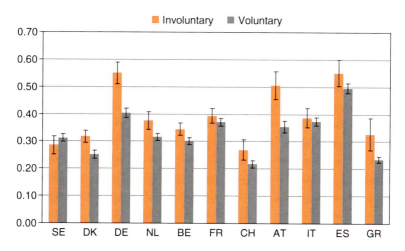

Fig. 17.1 Fraction reporting poor or fair health by job change status

While the substantial differences in reported health across countries also depend on the reporting style (Jürges 2007), the differences within a country show an interesting variation over Europe: in Sweden, for example, the difference in health between voluntary and involuntary job changers is basically non-existent, whereas in Germany or Austria, those who suffer from involuntary job change are much more likely to report bad health. This variation in within-country differences may be partially driven by how well the institutional settings (i.e. the respective unemployment benefit system) are able to help the individual when coping with unemployment.

17.4 The Analysis of Job Loss and Health

Before turning to the actual analysis of the effects of job loss on health, we first have to consider how involuntary job loss is related to health prior to that event – the approach Strully (2009) used as evidence that job displacement is exogenous to health. Childhood health is used here as a proxy for health prior to job loss. The bars in Fig. 17.2 represent regressions of the likelihood of involuntary job loss on an indicator of childhood health, thus comparing the two groups with involuntary job loss separately to those who always changed their job on their own account. Regressions with robust standard errors were run separately for men and women, the error bars reflect the 95% confidence intervals around the coefficients of health on job loss.

Even though childhood health is a retrospective measurement and thus clearly influenced by the respondents' current situation, the relationship shown in Fig. 17.2 is similar to what Strully finds in her analysis with a clear difference in the two groups: there are no effects of childhood health on the likelihood of being in the

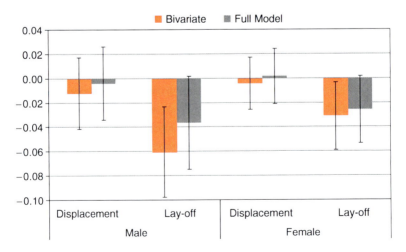

Fig. 17.2 Childhood health in relation to job change reasons

group of those suffering from plant closures (group 3), while reporting good or better childhood health reduces the likelihood of being in the group of those suffering from layoff or firing significantly (10% level). Although the effects are slightly smaller for women, there is no difference in the qualitative results.

The analysis now turns to the issue of displacement and health in the current situation. As mentioned above, we will look at three different health indicators: current self perceived health status (1 if fair or poor), chronic diseases (1 if at least one disease), and depression (1 if at least one symptom mentioned). We are interested in how these health indicators vary with the job change history – how do those who are displaced or fired fare compared to those who never had this experience? Figure 17.3 shows the coefficients from regressions run separately by gender of the respective health variable on the indicators for displacement (group 3 above) and lay-off (group 2), controlling for the wide range of variables mentioned in Sect. 17.2, i.e. demographic, childhood and parental characteristics as well as industry, occupation and country indicators. Hence we will tease out the differential effect of involuntary job loss by group compared to those never suffering from involuntary job loss. (The bivariate regressions did not lead to qualitatively different results and are thus not reported here.) The error bars again reflect the 95% confidence interval around the coefficients from a regression with robust standard errors.

The essential message in Fig. 17.3 is that both groups suffering from involuntary job loss are in worse health compared to those who never experience a displacement or a lay-off. However, one can also see small differences between the groups: those who were displaced show smaller effects, which also are not always significant at the 95% level. This is in line with what was postulated before – health and lay-offs may be intertwined, such that bad health causes the lay-off initially at least for some cases, which leads to larger effects here. Displacement does not seem to relate to

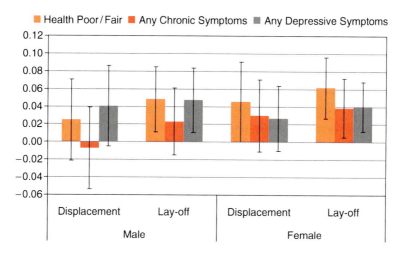

Fig. 17.3 Current health in relation to involuntary job loss

chronic diseases for men, while for females having suffered from lay-offs has a positive correlation with having chronic diseases. While females have slightly stronger effects on self reported health and chronic diseases, men seem to suffer more from depressions following an involuntary job loss. These gender differences are not significant, however.

17.5 Influences of the Welfare State

We now turn to the analysis of how the welfare state structure affects the current health of those individuals who have either been displaced or fired. We have seen the differences in health reports between involuntary and voluntary job changers in Fig. 17.1, and those differences across countries could have their reason in the institutional backgrounds of each country. Table 17.1 shows the current (2007) rules of unemployment benefits, where we specifically consider the replacement rate measured as the ratio to previous income, the duration of benefit payments, the qualifying conditions measured as the ratio of weeks an individual has to contribute to the system over the past year, and lastly the employee contribution as a percentage of income. (This table is a strong simplification of the reality in an attempt to put all countries into a common framework. The calculation of duration of benefits is partly based on weeks and thus produces "odd" numbers.) When looking at the welfare state variables, a clear-cut prediction of how they should influence the health of the unemployed is not clear – a more lenient welfare system could produce better health, but also in the long run it could lead to individuals less satisfied with their health.

The regression analysis using the unemployment benefit structure assumes that the combination of these measures is meant to help the individual reduce the negative effects of unemployment and thus aims at assessing the joint effect of these measures. For the purpose of evaluation, we now leave out the country fixed effects in the following regressions, and instead include the welfare state variables. However, all

Table 17.1 Unemployment benefit system variables by country

	Replacement rate	Duration of benefits (months)	Qualifying ratio	Employee contribution (%)
SE	0.70	9.9	1	0.00
DK	0.90	12.0	17	8.00
DE	0.67	12.0	17	3.25
NL	0.70	24.0	35	0.00
BE	0.60	12.0	30	0.87
FR	0.66	40.0	14	2.00
CH	0.70	13.2	26	1.00
AT	0.55	12.0	28	3.00
IT	0.30	7.0	26	0.00
ES	0.70	6.0	9	1.55
GR	0.40	8.3	16	1.33

other control variables mentioned above are still included in the model, which again is estimated with robust standard errors.

Since the previous analyses have shown that the differences between those displaced and those suffering from lay-off are small, we now return to the beginning and split the sample into those with an involuntary job change and those who changed voluntarily, i.e. group 1 compared to the combination of groups 2 and 3 (a separate regression did not yield different results). Each of the four welfare state variables in Table 17.1 are considered as explanatory variables as well as their differential influence (via interaction terms) on those who lost their jobs involuntarily. For reasons of brevity, the only dependent variable considered here is self-reported health. However, we calculate the effects separately for men and women. In order to elicit the effect of the welfare state on health for those who have lost their job, we use the coefficients of the interaction terms, calculate the effect of each welfare state variable at its respective country mean and then sum over all effects. Hence, we calculate the joint differential effect of replacement rate, duration of benefits, qualifying ratio, and employee contributions on the health of those suffering from involuntary job loss.

Results are shown in Fig. 17.4 separately for men and women, where each bar shows the joint differential effect of the welfare state variables on the likelihood to be in bad health for those who report having had any involuntary job loss. In both regressions, the hypothesis that the welfare state variables do not have any explanatory power has to be rejected (at 5% level), however, the joint *differential* effect is only significant for females. This is reflected in Fig. 17.4, showing much larger effects for women reporting involuntary job loss than men. The effect is clearly positive for women, as the likelihood to be in bad health is reduced in every country except Denmark by at least 10%. For men, only in Denmark we see an effect that is estimated to be larger than a 5% reduction in the likelihood to be in bad health. The gender difference may be in part a (positive) selection effect for the period under

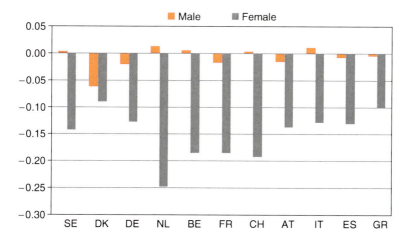

Fig. 17.4 Differential effect of unemployment benefits on health, by gender

consideration: while men were in general the main breadwinners and thus had to stay in the labour market, women who lost their job involuntarily and stayed in the labour market despite possible outside options such as marriage or childbirth, may have been in better health. However, since we control for childhood health, this effect would have to be rather strong.

17.6 Summary

This paper shows that there are long term effects of involuntary job changes on health. Both individuals reporting that they were laid off and those reporting to have lost their jobs due to plant closures are in worse health than those individuals never experiencing lay-offs or displacement. This holds true for three different health indicators: self-perceived health, chronic diseases, and depressive symptoms. It is likely, that the sample analysed here suffers from a survivor bias: individuals who have suffered greatly from a job loss may already have died. However, this would bias the results towards zero, such that the observed effects may in reality be even stronger.

Men and women are almost equally affected by an involuntary job loss in terms of health, although women experience slightly stronger effects. In part, this may be due to a selection into the workforce – women in the age group considered may have had substantially different motives to work, which the control variables are not able to pick up.

The differential effects of the unemployment benefit system on the health of those with involuntary job loss have to be considered with care, because of the simplifications mentioned and also because their inclusion may also pick up pure state variation. However, the difference between men and women is of interest – controlling for age, occupation, industry and socio-demographic background variables, it seems as if the welfare state has particularly helped women in reducing negative effects of involuntary job loss on health.

References

Bono, E. Del., Weber, A., & Winter-Ebmer, R. (2008). *Clash of career and family: Fertility decisions after job displacement.* Centre for Economic Policy Research, Discussion Paper No. 6719, London.

Browning, M., Dano, A. M., & Heinesen, E. (2006). Job displacement and stress-related health outcomes. *Health Economics, 15*, 1061–1075.

Charles, K. K., & Melvin, S., Jr. (2004). Job displacement, disability, and divorce. *Journal of Labor Economics, 22*(2), 489–522.

Eliason, M., & Storrie, D. (2009). Does job loss shorten life? *The Journal of Human Resources, 44*(2), 277–302.

Jacobson, L., LaLonde, R., & Sullivan, D. (1993). Earnings losses of displaced workers. *American Economic Review, 83*(4), 685–709.

Jürges, H. (2007). True health vs. response styles: Exploring cross-country differences in self-reported health. *Health Economics, 16*(2), 163–178.

Ruhm, C. (1991). Are workers permanently scarred by job displacement? *American Economic Review, 81*(1), 319–324.

Salm, M. (2009). Does job loss cause ill health? *Health Economics, 18*, 1075–1089.

Strully, K. W. (2009). Job loss and health in the U.S. labor market. *Demography, 46*(2), 221–246.

Sullivan, D., & von Wachter, T. (2009). Job displacement and mortality: An analysis using administrative data. *The Quarterly Journal of Economics, 124*(3), 1265–1306.

Chapter 18
Life-Course Health and Labour Market Exit in 13 European Countries: Results From SHARELIFE

Mauricio Avendano and Johan P. Mackenbach

18.1 Health and Labour Market Participation

Despite dramatic increases in life expectancy during the second half of the twentieth century, the proportion of lifetime individuals spend in the labour force has decreased in most European countries during the last decade. Although the official retirement age in most of Europe ranges between 60 and 65 for women and 62 and 65 for men, the age at which Europeans actually withdraw from the labour market varies markedly across countries, and is generally well below the official age (Romans 2007). Among women, the median retirement age ranges from 55 in Poland and Slovenia, to 63 in Sweden; among men, it ranges from 57 in Poland, to 65 in Estonia and Cyprus (Romans 2007). The determinants of these variations are not well understood due to the complexity of retirement decisions and pathways to retirement. Understanding the determinants of these variations is essential for the development of policies that can increase labour force participation in Europe.

Particularly at old age, health is a major determinant of transitions in and out of employment (Haan and Myck 2009). The mechanisms through which illness may lead to labour market exit can be conceptualized in two ways: First, illness may lead to impairments that limit the ability to work; second, illness may lead to eligibility to public disability benefit programmes, which enable individuals to earn non-wage income while out of the labour market. Consistent with these two mechanisms, two types of policies may influence the rate of exit from the labour market as a result of illness. First, health promotion and prevention policies have the potential to produce healthy workers that remain longer in the labour force. Second, higher generosity of

---

M. Avendano (✉)
Department of Public Health, Erasmus MC, room AE-104, P.O. Box 2040, 3000 CA Rotterdam, Netherlands
e-mail: mavendan@hsph.harvard.edu

J.P. Mackenbach
Department of Public Health, Erasmus MC, University Medical Center Rotterdam, room number Ae-228, P.O. Box 2040, 3000 CA Rotterdam, The Netherlands
e-mail: j.mackenbach@erasmusmc.nl

public disability benefit programmes might be an incentive for disability insurance uptake, leading to higher exit rates attributable to illness (Börsch-Supan 2006, and Chap. 19). Policies on these domains differ dramatically across countries and may influence the link between illness and labour force participation. Cross-national comparisons provide a unique opportunity to identify policies that can ameliorate the impact of health on labour force participation in Europe.

Previous studies have focused primarily on the short-term impact of mid- and old-life health on retirement (Bound et al. 1998; Disney et al. 2006; Dwyer and Mitchell 1999; Haan and Myck 2009; Romans 2007; Tanner 1998), but less is known about how health since the early years of productive age influences the likelihood of leaving the labour force later in life. In this paper, we examine the impact of periods of long-term illness over the life-course on labour force participation in a sample of 13 European countries participating in SHARELIFE. Three hypotheses underlie our analysis: First, we hypothesize that periods of illness over the life cycle have a long-standing impact on the ability to work and the likelihood of leaving the labour market at older ages. Second, we expect larger government investments in public health to reduce the impact of illness on the ability to work, thus leading to gains in economic productivity. Finally, we hypothesize that more generous disability insurance and unemployment benefit programmes lead to a stronger association between health and labour market exit, as these may work as financial incentive towards earlier health-related exit from the labour market.

18.2 Measuring Labour Market Exits and Health

Our analysis is based on men who participated in the SHARELIFE survey, and who were also interviewed in the first wave of SHARE. Due to the complexities of labour market histories in older cohorts of women, our analysis focuses on men only. Data from SHARELIFE was linked to basic demographic and labour force participation data from the first wave (2004/2005) of SHARE, except for the Czech Republic and Poland, where the first wave took place in 2006/2007.

Based on the life History Event Calendar approach we reconstructed life histories of adult health using the questions asking participants about the number of periods of ill health or disability experienced during adulthood that lasted more than a year. Individuals were asked the starting and ending year of each period of illness, as well as the specific medical condition that accounted for each period.

For work participation, individuals were asked about their entire labour market history, including transitions in and out of employment over the life-course. Participants reported the date on which they entered the labour market, at which point follow-up started. Based on the approach of previous studies (Disney et al. 2006), labour market exit was defined as the last observed exit from economic activity (if observed), thus assuming inactivity as an absorbing state.

We exploit the event history structure of the data to assess long-term effects of health on labour force participation in failure-time or duration models (Box-Steffensmeier and Jones 2004). We reconstructed the date at which individuals entered the labour market and observed individuals thereafter. During the observation period, individuals were at risk of experiencing a transition from employment to non-employment. Individuals that did not experience a transition out of employment were censored, because although the outcome event may be experienced, employment status after the last point of observation was unobserved.

The Cox Proportional hazard model was used to model survival time in the labour market as a function of illness status, controlling for age of entry into the labour market, education, age of illness and country. We are interested in the effect of experiencing at least one period of illness during adulthood on "survival" in the labour force. Illness status and age of illness were treated as time-varying covariates. Individuals could change the value of illness status from "no illness" to "illness". Similarly, individuals and periods without illness were assigned the mean age of illness, which changed to the observed age of first illness for those who reported having had an illness period.

Analysis was restricted to individuals who at some point in their lives entered the labour market, because those who did not enter employment were not at risk of experiencing a transition out of the labour force. Thus, individuals who reported that they had never done any paid work were excluded. The final sample size included 8,132 male participants. Individuals who started to work immediately after completing school were assigned the date in which they finished high school as the date of entry into the study. Individuals that reported a gap of non-employment after high-school were assigned the date of their first job as the date of entry into the study. Participants who started to work before completing their full-time education did not report the exact age and were assigned the median age of entrance into the labour market of other participants (18 years). Individuals who were still employed at the time of interview were right-censored at interview date.

18.3 Illness over the Life-Course

Figure 18.1 shows the age-adjusted prevalence of reporting one or more periods of illness that lasted at least 1 year during the years of productive life for the 13 European countries in SHARELIFE. Overall, 13% of participants reported having experienced an episode of illness, but there were large variations across countries. Switzerland (8%) and France (9%) experienced the lowest prevalence of illness, while the highest prevalence was observed in Poland (17%) and the Czech Republic (27%).

The nature of these periods of illness is summarized for the entire sample of males in Fig. 18.2. Percentages add to more than a hundred because individuals could report more than one condition as responsible for each period of illness. Heart disease,

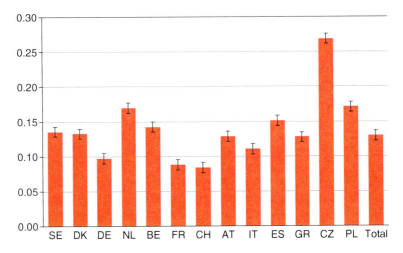

Fig. 18.1 Prevalence of reporting one or more periods of illness before labour market exit in men at ages 50 years and older

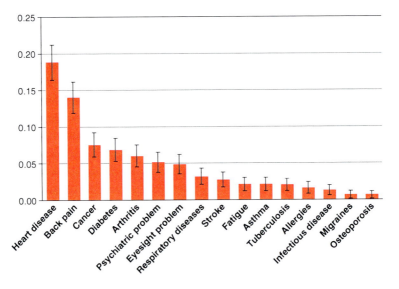

Fig. 18.2 Medical conditions that account for the first period of illness in men aged 50 years or older who experienced at least one illness period

including angina, heart attack and other heart diseases, accounted for the largest share of periods of illness lasting at least 1 year (19%). The second major cause of periods of illness was back pain (14%), followed by cancer (7%), diabetes (6%), arthritis (6%), psychiatric conditions (5%), and eyesight problems (5%). Other listed conditions accounted each for less than 4% of periods of long-term illness. In total, 30% of illness periods were attributable to conditions that were not included in the list.

The distribution of conditions was overall very similar across all countries. However, there were some important regional differences. In the Nordic countries as well as in Western Europe, around 20% of all cases of illness was attributable to back pain, compared to only 7% in Southern Europe and 10% in Eastern European countries. While heart disease and angina accounted for 8% of illness periods in the Nordic countries and 14% in Western European countries, this figure was 23% for Southern Europe and 24% for Eastern European countries. Cancer had a larger share of all illness in Nordic (9%) and Western (10%) European countries than in Southern (5%) and Eastern (6%) European countries. About 20% of all periods of illness in Northern Europe was attributable to other conditions not included in the list, compared to around 30–33% in Western, Southern or Eastern European countries.

18.4 The Impact of Illness on Labour Market Exit

For a selected set of countries, Fig. 18.3 shows survival curves that summarize the length of time individuals survive in the labour market before exiting (or censoring), according to whether they experienced at least a period of illness earlier in their life. Participants who never experienced a period of illness ("healthy life") had generally higher rates of labour force participation and lower exiting rates at all ages than individuals who at some point in their life experienced a period of illness. Although in some countries the lines approached each other, in most countries differences were substantial. For example, in Austria, at age 55, 74% of men who had never experienced a major illness were still in the labour market, as opposed to only 57% of those who had experienced a period of long-term illness. By age 60, 40% of those in good health were still in the labour market, as opposed to only 18% of those who had a period of illness. In France, 72% of healthy men were still in the labour force at age 55, compared to 63% of those who had experienced a period of illness. At age 60, 37% of healthy men were in the labour market, compared to only 24% of those who had experienced a period of illness. In Italy, as well as in Greece and Spain, there was little difference in the employment trajectories of those with and without a previous history of long-term illness. However, survival curves ignore the time-varying nature of illness, underestimating potential differences in survival.

Figure 18.4 shows the hazard ratio of labour market exit according to whether individuals had experienced at least one period of illness lasting a year or more over their life course, incorporating illness as a time-varying covariate. Effects appear now to be much more consistent across countries. In the total sample, a period of long-term illness is significantly associated with twice higher risk of exiting the labour market over the years of productive life. There were, however, important

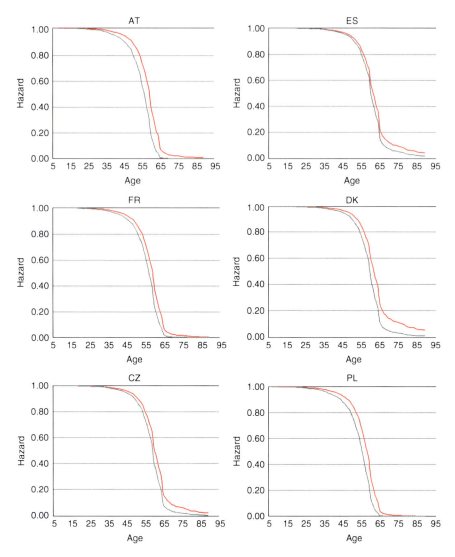

Fig. 18.3 Survival in the labour market according to age, split by illness status in a Cox Proportional Hazard model for men at ages 50 years and older

variations across countries. Long-term illness increased the risk of future labour market exit by almost three times in Germany and Poland, and by around twice in Austria, Denmark, the Czech Republic and Spain. Strong effects were also observed in Belgium, the Netherlands, France, Italy and Sweden, where illness increased the risk of labour market exit by 60–80%. Effects were not significant in Switzerland possibly due to the small sample size, while in Greece, there was no evidence of an association.

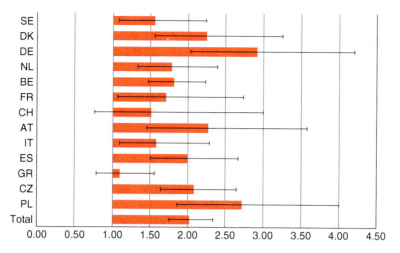

Fig. 18.4 Hazard ratio of last labour market exit according to illness status in a time-varying Cox Proportional Hazard model for men at ages 50 years and older

18.5 The Role of Public Health and Labour Policies

Differences among countries in the impact of illness on labour market exit may be explained by a variety of both health care and labour policies. In this section, we explore to what extent expenditures in relevant policies are correlated with life-course health and the magnitude of the association between illness and labour market exit.

Panels A to D in Fig. 18.5 summarize the association of expenditures in public health and curative care as percentage of GDP with the prevalence of long-term illness and the hazard ratio that summarizes the association between illness and labour market exit. Panels A and B show that expenditures in public health were not correlated with the prevalence of long-term illness or the association between illness and labour market exit. In contrast, Panels C and D show that higher expenses in curative care were associated with lower prevalence of long-term illness ($r = -0.74$, $p = 0.015$), and hazard ratios on the impact of illness on labour market exit ($r=-0.44$, $p = 19$), although only the former was significant.

Panels E and F show that while there was no significant correlation for expenditures in disability and sickness ($r = 0.22$, $p = 47$), higher expenditure in unemployment benefit programmes was associated with a stronger association between illness and the risk of labour market exit ($r = 0.60$, $p = 0.053$). In supplementary analyses, we explored this further by examining correlations with unemployment benefit duration and average net replacement rates for the year 2004 from the OECD employment outlook. There was a non-significant positive correlation between unemployment benefit duration and the impact of illness on labour market exit ($r = 0.329$, $p = 0.324$). In contrast, although the correlation was not significant, higher average net replacement rates over 60 months of unemployment seemed to be correlated with a stronger association between illness and labour market exit ($r = 0.46$, $p = 0.11$).

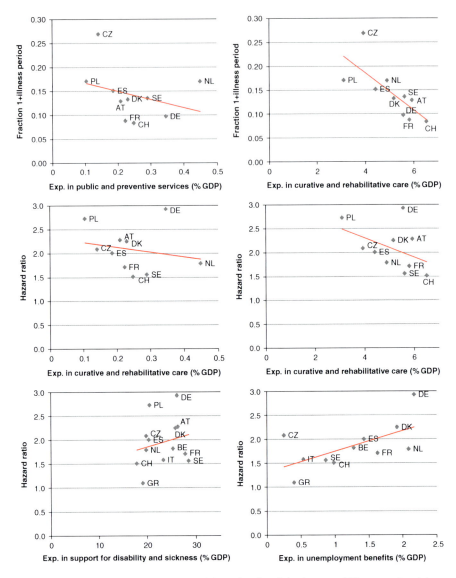

Fig. 18.5 Welfare state interventions and hazard ratio of the impact of illness on last labour market exit in men at ages 50 years and older

18.6 Conclusions

Three hypotheses guided our analysis. First, we hypothesized that periods of illness over the life cycle have a long-standing impact on the ability to work and the likelihood of leaving the labour market at older ages. In support of this hypothesis, we found that in most European countries, long-term illness is associated with earlier exit from the labour market. We hypothesized that higher public health investments might ameliorate this association. Our results are inconclusive but generally do not suggest a strong correlation of the level of public health investments with the prevalence of long-term illness or the magnitude of the association between illness and labour force participation. Investments in curative health care do seem to be strongly associated with the prevalence of long-term illness and the impact of illness on labour market exit, although correlations were not significant for the latter. Our third hypothesis referred to the role of labour market programmes. We found that larger investments in unemployment benefit programmes are associated with a larger impact of illness on labour force participation. Although correlations with average replacement rates and unemployment duration policies did not reach statistical significance, overall findings suggest that higher unemployment benefits may potentially work as incentive towards earlier exit from the labour market due to illness. (These findings are confirmed by Börsch-Supan and Roth, Chap. 19).

It is important to consider a number of limitations in our study. First, our analysis examines associations, and ignores the fact that illness might be endogenous to labour force participation. Lower labour force participation may lead to poorer self-reports of health, or both health and labour force participation may be the result of unobserved characteristics such as parental investments or family background. Therefore, although our results are consistent the hypothesis that poor health leads to shorter working lives, they are also consistent with other hypotheses involving reverse causality or unobserved heterogeneity. Second, our study is based on a life-history event approach, and may therefore suffer from mortality selection, as individuals who did not survive to become part of the survey are unobserved. If poor health increases the risk of exiting employment, we would expect mortality selection to downwardly bias estimates, as those in poor health would be more likely to be unobserved. Third, information on periods of illness and employment histories might be susceptible to recall bias. Finally, our assessment of the impact of welfare state policies is only a first approximation that ignores that time-varying nature of employment and public health policies. Future analyses should focus on linking national data on policies over time to the particular timing of events of illness and job loss experienced by participants.

Despite these limitations, our study raises important questions regarding the impact of poor health on work and the potential mechanisms behind this association. Findings are consistent with previous studies showing that adverse shocks to individual health predict individual retirement (Disney et al. 2006; Haan and Myck 2009). Our results extend these findings by suggesting that periods of illness impact duration in the labour market, even if these events occurred at a relatively early point in the life-course. Previous evidence suggests that ill health is a major reason for retirement among men in several European countries (Bound et al. 1998; Tanner 1998). It has been suggested that this association may be explained by the "justification hypothesis", i.e., self-reports of health are misreported as individuals use health as justification for leaving the labour force (Bound 1991). This phenomenon is unlikely to explain our results, because periods of illness assessed in our study referred often to health shocks that occurred many years before retirement decisions took place. In addition, our measure is not based on an overall self-assessment of health, but on self-reports of specific periods of illness accounted by specific medical conditions.

Several plausible mechanisms may explain the association between illnesses and labour market exit. Poor health may directly prevent productive activity thus leading to early retirement. Periods of long-term illness may be indicators for general poor health, so that individuals who experience a period of illness earlier in life may be prone to experience late-life problems that ultimately lead to early retirement. Poor health may also lead to a change in preferences and the "price" of leisure time, as the need for non-work time to care for health increases. Our results are consistent with these two potential mechanisms linking health and earlier exit from the labour market. On the other hand, poor health may also lead to higher consumption of health care, potentially forcing individuals to delay retirement to meet their consumption needs. This second explanation is not consistent with the finding that ill health leads to earlier exit from the labour market. This may be due to the fact that European countries enjoy close to universal health care coverage, potentially diminishing the need to increase work to meet health care consumption needs. Overall, our results are consistent with the view that poorer health leads to earlier exit from the labour market because of its effects on preferences and productivity (Dwyer and Mitchell 1999).

Our results suggest that the association between poor health and labour market exit is complex and varies substantially across countries with different welfare state policies. We find that higher investments in unemployment benefit programmes are associated with stronger associations between ill health and retirement. This suggests that while health as such plays an important role, individuals also respond to incentives to retirement, and may use unemployment benefit programmes as a potential pathway to leave the labour market as a result of illness.

In conclusion, our study suggests that periods of illness during the life-course are associated with an increased risk of leaving the labour force, but the magnitude of this association varies across countries. Our results suggest a new area of exploration to disentangle how historical reforms to labour market programmes have influenced the role of illness in early retirement, and how this varies across European countries with different welfare state policies and institutions.

References

Börsch-Supan, A. (2006). Work disability and health. In A. Börsch-Supan (Ed.), *Health, ageing and retirement in Europe: First results from the survey of health, ageing and retirement in Europe* (pp. 253–258). Mannheim: Mannheim Research Institute for the Economics of Aging (MEA).

Bound, J. (1991). Self-reported versus objective measures of health in retirement models. *Journal of Human Resources, 26*, 106–138.

Bound, J., Schoenbaum, M., Stinebrickner, T. R., & Waidmann, T. (1998). *The dynamic effects of health on the labor force transitions of older workers.* NBER working paper w6777, National Bureau of Economic Research, Cambridge, MA.

Box-Steffensmeier, J., & Jones, B. (2004). *Event history modeling: A guide for social scientists.* Cambridge: Cambridge University Press.

Disney, R., Emmerson, C., & Wakefield, M. (2006). Ill health and retirement in Britain: A panel data-based analysis. *Journal of Health Economics, 25,* 621–649.

Dwyer, D., & Mitchell, O. (1999). Health problems as determinants of retirement: Are self-rated measures endogenous? *Journal of Health Economics, 18,* 173–193.

Haan, P., & Myck, M. (2009). Dynamics of health and labor market risks. *Journal of Health Economics, 28*(6), 1116–1125.

Romans, F. (2007). *The transition of women and men from work to retirement.* Brussels: European Communities.

Tanner, S. (1998). The dynamics of male retirement behaviour. *Fiscal Studies, 19,* 175–196.

Chapter 19
Work Disability and Health Over the Life Course

Axel Börsch-Supan and Henning Roth

19.1 Work Disability Across European Countries

Disability insurance – the insurance against the loss of the ability to work – is a substantial part of social security expenditures and an important part of the welfare state regime in all developed countries (Aarts et al. 1996). Like almost all elements of modern social security systems, disability insurance faces a trade-off. On the one hand, disability insurance protects unhealthy people who are not able to work from falling into poverty before they are eligible for normal retirement benefits. On the other hand, however, disability insurance creates incentives to exit the labour force early and may act as another pathway to early retirement without the incidence of a major health loss.

The recipiency rates of disability insurance (DI) benefits vary strikingly across European countries, see Fig. 19.1. They are defined as the share of all individuals aged 50–64 who receive benefits from DI. With 15.6% and 11.6% the Nordic countries Sweden and Denmark have fairly high recipiency rates. The Central European countries cover a broader range. The rate of the Netherlands is 14.0% and thus similar to the Nordic countries while in France only a 1.7% of the people receive DI benefits. In the Mediterranean countries lower rates can be observed varying from 3.3% in Greece to 9.0% in Spain. The Eastern European countries exhibit the highest recipiency rates. While the Czech Republic with 12.3% is in a range with Denmark or the Netherlands, the Polish rate of 19.2% exceeds the rest by far.

Why are so many more individuals aged 50–64 receiving DI benefits in Denmark, Sweden and the Netherlands than in e.g. France or Germany? Why so many fewer individuals in Greece than in Poland? This chapter investigates the causes for this variation. Three candidate causes come to mind: cross-national differences in the age structure, cross-national differences in health, and cross-national differences in the early retirement incentives created by the DI system. In earlier work

---

A. Börsch-Supan (✉) and H. Roth
Mannheim Research Institute for the Economics of Aging (MEA), L13, 17, 68131 Mannheim, Germany
e-mail: axel@boersch-supan.de; roth@mea.uni-mannheim.de

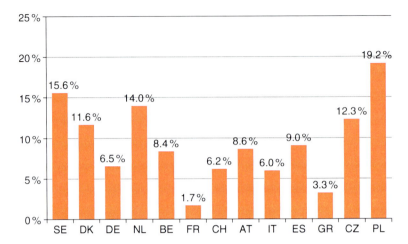

Fig. 19.1 DI recipiency rate in 13 European countries
Source: SHARE wave 2 (2007), population weighted data

based on the 2004 waves of SHARE, ELSA and HRS, we showed that cross-sectional differences in demographic structure and current health status cannot explain the cross-national differences in DI recipiency (Börsch-Supan 2005) although health explains a great deal of the within-country variation (see also Chap. 18). A second stage of our research was based on two waves of data. We showed that health events between waves did not significantly trigger a higher probability of becoming a DI benefit recipient (Börsch-Supan 2008).

The poor explanatory power of a broad battery of health measures used in these studies, including objective and subjective measures as well as performance measures of physical and mental health, is disturbing and undermines the role of DI as an insurance of last resort against failures of health in working age. It has been criticized, however, that current health measures, as broad as they may be measured, do not appropriately capture the full impact of poor health on employability. Rather, it is argued, work disability is the result of a long lasting process of becoming sick and finally unable to work.

This paper therefore takes a life-course approach. Thanks to the new SHARE-LIFE dataset, we are able to add to the analysis a set of variables that account for those long-run effects. We first create lifetime health indicators that describe childhood and adulthood health status. In addition, we take other life-course features into account such as childhood socio-economic status, quality of the working place and marital status over the whole life course.

In the following section, we will briefly describe our approach. We then present our results at the individual level. We find that both current and life-course health significantly influences the probability of receiving DI benefits. We then turn to the cross-national level. We find that welfare state differences dominate at this level while cross-national health differences remain largely irrelevant even when taking life-course health measures into account.

19.2 Variables and Technique

We focus our analysis on people at the age between 50 and 64 because this is the time span in which exiting the labour market via DI may be an attractive opportunity for early retirement. Beginning with age 65, normal retirement benefits are available in all 13 countries in our analysis. The baseline of analysis is the year 2007. We have a large number of 10,385 observations, on average 800 in each country, with substantial differences across countries. Our dependent dummy variable is the recipiency of DI benefits. Following the three candidate causes and distinguishing current status from life-time influences, we employ five categories of variables:

1. Current basic demographic characteristics: age, gender and years of education.
2. A broad range of variables describing current health: self-perceived health, functional physical status described by the number of limitations in activities of daily living (ADL) and limitations of instrumental activities of daily living (IADL), mental health status as measured by EURO-D, grip strength as indicator of physical performance.
3. Life course health indicators include childhood health status and adulthood health status. Childhood health is described by the number of illnesses lived through until the age of 15. For adulthood health a similar measure is taken, and in addition a binary variable indicating if someone had suffered from an extended period of poor health. Moreover, we include the number of gaps in the working history in which a person was sick or disabled.
4. Life course control variables include childhood socio-economic status, work quality and marital status. The socio-economic status during childhood is measured by the number of books, rooms per person in the accommodation and relative skills in mathematics at the age of 10. Work quality is measured as the subjective assessment of the physical and psychological demands at work. We also account for the number of jobs during lifetime. Finally, we include binary variables indicating if someone has been married, divorced or widowed during her or his lifetime.
5. Variables describing the generosity of the welfare system regarding DI and alternative pathways are taken initially from OECD (2003). We have updated and extended these indicators to the countries not covered by the OECD. In general the OECD gives scores from 0 to 5 whereat a higher score represents a more generous system. At DI coverage 5 points are given if the DI covers the whole population while 0 points represents coverage only for employees. The minimum disability level that is required to be eligible is measured as percentage measure of work disability. The lower the percentage required the higher the

score given. The maximum benefit level is measured as a replacement rate. A higher rate leads to a higher score. The strictness and whether DI benefit eligibility requires a medical assessment or whether a vocational assessment is sufficient is also included in the analysis. Finally, we insert a measure for the strictness of the unemployment insurance as an alternative pathway of early retirement.

Our analysis is divided into two parts. First, we relate at the individual level whether a person receives DI benefits to the above set of explanatory variables. We do this by pooling the SHARELIFE data from all countries and performing probit, logit and linear regression analyses. We also assess how much total variation in DI benefit recipiency at the individual level is explained by the different categories of variables.

Second, we analyze the cross-national variation depicted in Fig. 19.1. To do so, we perform simulations which hold some of the explanatory variables counterfactually constant. If this group of variables were the main cause for the international variation, the simulated outcome should produce roughly identical percentages of DI benefit recipiency in each country.

19.3 Regression Results at the Individual Level

Since the dependent variable – receipt of DI benefits – is binary we begin with a probit and a logit specification. Only the probit results are shown below since they yield very similar results explaining about 23% of the total variation (measured as the pseudo R^2) which is quite a satisfactory value at the individual level. We also used a linear specification because it delivers essentially the same regression results (although on a different scale) and permits a more straightforward way to decompose the total variance.

All five categories of variables are jointly statistical significant: the corresponding F-test values are 23.4 for demographic variables, 208.6 for current health measure, 29.6 for the welfare state indicators, 201.9 for life-course health and 90.3 for all other life-course variables. Table 19.1 presents the results for the probit and linear specification. For the probit model, marginal effects are shown rather than the regression coefficients.

Age and years of education have a negative effect on the receipt of DI benefits. Hence, older individuals have a smaller probability of receiving DI benefits. This may sound counterintuitive since health declines as we age. However, we control for health, see below, and alternative retirement pathways become available at older ages. More educated individuals are less likely to receive DI benefits. Male individuals are more frequently DI benefit recipients than female.

All current health measures have the expected sign and are significant, except for the number of ADL limitations. A dummy variable of the presence of ADL limitations, however, is significant. Better health leads to a lower probability

Table 19.1 Determinants of DI recipiency

Variables	Probit Marginal effects	Standard error	Linear Coefficients	Standard error
Age (years)	−0.001	(0.0006)	−0.001**	(0.0006)
Gender (dummy)	−0.044***	(0.0070)	−0.056***	(0.0082)
Education (years)	−0.002***	(0.0007)	−0.002**	(0.0007)
Self-perceived health (1–5)	0.038***	(0.0028)	0.039***	(0.0030)
ADL (0–6)	0.002	(0.0050)	0.022	(0.0140)
IADL (0–7)	0.025***	(0.0046)	0.067***	(0.0125)
Maximal grip strength (kg)	−0.001***	(0.0003)	−0.001***	(0.0004)
EURO-D (0–12)	0.004***	(0.0012)	0.005***	(0.0016)
Childhood illnesses (0–7)	−0.003	(0.0027)	−0.005	(0.0031)
Adulthood illnesses (0–5)	0.017***	(0.0026)	0.037***	(0.0058)
Working gaps due to sickness (0–2)	0.052***	(0.0114)	0.118***	(0.0301)
Period of very poor Health (dummy)	0.056***	(0.0051)	0.060***	(0.0061)
Rooms per person	−0.003	(0.0065)	−0.002	(0.0030)
Number of books (dummy)	−0.002	(0.0055)	−0.002	(0.0056)
Mathematical skills (dummy)	−0.007	(0.0052)	−0.005	(0.0050)
Number of jobs	−0.003***	(0.0013)	−0.005***	(0.0013)
Physical demand of work (dummy)	0.022***	(0.0053)	0.024***	(0.0062)
Psychological demand of work (dummy)	−0.005	(0.0049)	−0.007	(0.0051)
Married (dummy)	−0.013	(0.0089)	−0.019*	(0.0105)
Divorced (dummy)	0.012*	(0.0063)	0.015**	(0.0072)
Widowed (dummy)	0.005	(0.0094)	0.005	(0.0109)
Coverage (0–5)	0.010***	(0.0030)	0.011***	(0.0036)
Minimum disability level (0–5)	0.010***	(0.0027)	0.009***	(0.0025)
Replacement rate (0–5)	−0.007**	(0.0029)	−0.006**	(0.0027)
Medical assessment (0–5)	0.005*	(0.0025)	0.007***	(0.0028)
Vocational assessment (0–5)	−0.017***	(0.0028)	−0.017***	(0.0034)
Unemployment benefits (0–5)	0.013***	(0.0043)	0.014**	(0.0055)
Constant			0.125**	(0.0599)

***, **, *: Significant at 1%, 5%, 10%, respectively

of receiving DI benefits. As a remarkable result, we find that the more subjective a health measures is, the stronger is its influence. This may be an indication of some extent of self-justification (see Banks et al. 2004).

The life-course health variables show a clear picture. All life course indicators describing long-term health show the expected direction. Moreover, these variables are highly significant jointly but also each for itself as it can be seen in the table above. This result is robust over all three specifications. The variable describing childhood health is not significant.

Among the other life-course variables, the only significant ones are the subjective physical demand of work, the number of jobs and the binary variable describing if someone has been already divorced. Higher physical demand of the work leads to a higher probability of receiving benefits while an increase in the number of jobs leads to a decline in the reception of DI benefits. Suffering from at least one divorce increases the probability of being eligible.

The OECD indicators describing the generosity of the welfare system regarding DI and alternative pathways vary only across countries. They are nevertheless jointly significant and have, besides the replacement rate, the expected direction: the more generous the DI, the higher the probability of receiving the benefits. The broader the job range of vocational assessment, the less likely is the receipt of DI benefits. Strict eligibility rules and a low replacement rate of the unemployment insurance, a possible alternative pathway to retire early, increase the likelihood of receiving DI benefits.

Figure 19.2 shows how much variation at the individual level is explained by each of the five groups of variables, for simplicity using the linear regression model. The full linear model explains some 14.65% of the variation in the data. Basic demographic characteristics and education explain less than 1% of the individual variation. The OECD indicators vary only across countries and therefore explain, by definition, very little at the individual level. Current health measures have the largest explanatory power with over 9% of the individual variation explained. Life-course health variables are almost as powerful and explain 7.2% of the individual variation, while the other life-course variables explain 6.5%. These results are in line with the findings by Avendano and Mackenbach, Chap. 18.

Quite clearly, both current and life-course health are highly predictive of receiving DI benefits at the individual level. Together, the health variables explain 12.4% of the total variation, i.e. 85% of the explained variation. Self-rated health is far the strongest single health variable, explaining 6.8% of the total variation.

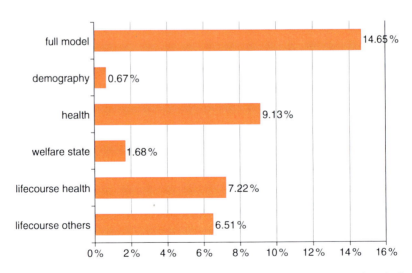

Fig. 19.2 Explanatory power of variable groups (in % of explained variation). Based on the linear regression model

19.4 Counterfactual Simulations at the Country Level

This decomposition is dominated by intra-country individual variation and therefore does not shed much light on the large variation across countries. In order to separate cross-national from within-country variation, we predict DI recipiency rates at the country level with a counterfactual simulation which sets potential explanatory variables at the same level (usually the sample average) for all individuals across all countries. If a group of variables were the main driver of cross-national differences of DI recipiency, then equalizing these variables should also equalize the DI recipiency rates.

We perform three sets of such counterfactual simulations. The first set reproduces the results of Börsch-Supan (2005) with the 2007 data. It equalizes the demographic structure (i.e., all individuals are counterfactually assigned the same age, gender and number of years in education) and health. Figure 19.3 shows the resulting counterfactual cross-national distribution of DI recipiency rates.

The second bar for each country in Fig. 19.3 equalizes age, gender and education across countries. Quite clearly, the resulting counterfactual DI benefit recipiency rates are virtually identical to the actual rates, represented by the left bar. Hence, age, gender, and education differences across countries can be ruled out as drivers of the cross-national variation in DI recipiency.

The right bar equalizes all current health measures. In countries with a relatively low level of health (especially Poland) and in Switzerland, where health is particularly high, we can indeed attribute some of the cross-national variation in DI benefit recipiency to health since the counterfactual prediction puts these countries closer to the average. The opposite, however, is true for those three countries in the EU15 in which benefit recipiency rates were particularly high in 2007: Sweden,

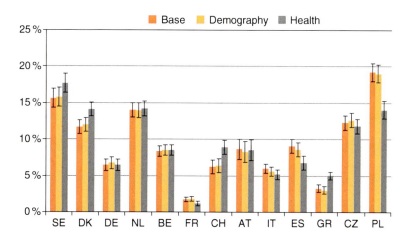

Fig. 19.3 Demography and health. Simulation based on linear regression model. *Brackets* denote standard errors

Denmark, and the Netherlands. For these countries, correcting for health exacerbates the cross-national differences rather than levelling them off. Moreover, e.g. in Germany, measurements of objective health turn out to be better than the European average, while self-rated health is reported to be lower than the European average. Equalizing both objectively measured and self-rated health thus compensates each other. The opposite can be observed in Belgium. In summary, except for Poland, Greece and Switzerland, current health is not a main driver of cross-national variation in DI recipiency. This is a remarkable result as DI recipiency should be linked to work disability and thus health.

Current health, however, even if broadly measured, may be too narrow a health measure to determine the probability of receiving DI benefits because health events which took place much earlier in life may have driven the transition out of work. Possible influences are multi-dimensional. There may be direct effects of childhood diseases that have undermined resilience in old age and then lead to a disability. There may also be indirect effects of childhood diseases that worsen adult health at earlier stages. Often, disabilities are the result of long periods of illness and suffering from physical or mental impairments. Current health measures cannot reflect such long-term developments. Moreover, there may be other childhood living conditions such as socio-economic status that may build the background for later health problems and disability.

We take account of these possibilities by performing a second set of counterfactual simulations, now equalizing the life-course health and other life-course variables available in the SHARELIFE data, such as indicators for socio-economic status, marital history and work satisfaction.

Figure 19.4 presents the results. They are unambiguous: life-course variables do not explain the cross-national variation. Not a single difference is statistically significant.

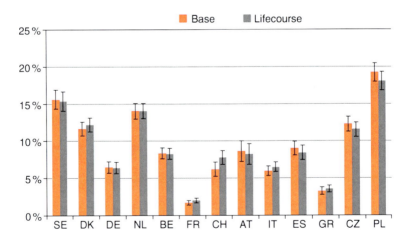

Fig. 19.4 Life-course health and other life-course factors. Based on the linear regression model. *Brackets* denote standard errors

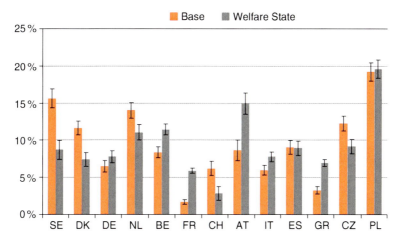

Fig. 19.5 Welfare state generosity. Simulation based on linear regression model. *Brackets* denote standard errors

So far, we have ruled out demographics, education, current and life-course health and other life-course characteristics as causes for the cross-national differences in DI benefit recipiency. Among the variables discussed in the introduction, institutional features and their incentives created remain as another potential cause. Our third set of counterfactual simulations therefore equalizes all variables that describe the generosity of the DI system and potential alternative pathways, such as unemployment insurance.

Figure 19.5 shows that actual and simulated now diverge considerably. Except for Switzerland and Poland, the simulated recipiency rates of DI benefits are much more equal across countries when we assume the same institutional framework in every country. Most importantly and as opposed to Fig. 19.3, those three countries in the EU15 in which benefit recipiency rates were particularly high in 2007 – Sweden, Denmark, and the Netherlands – now exhibit much smaller DI rates when the generosity of their DI systems is reduced to the average level across the 13 included countries.

19.5 Discussion and Conclusions

In assessing our results, it is important to distinguish individual level variation from cross-national variation. Since we have more than 10,000 observations and only 13 countries, our regression results (Table 19.1) are dominated by the within-country variation. Here, both current and life-course health variables are highly significant both jointly and each for itself at the individual level. This shows that these variables are reliable measures of health, and they indeed contribute to about 85% of the overall explained variation across individuals. Variables describing the welfare state, however, especially the generosity of the DI system, cannot determine within a country if someone receives DI benefits because all individuals face the same DI system.

In our counterfactual simulations (Figs. 19.3–19.5), we only see the cross-national variation. At this level, the roles of health and DI system generosity switch completely. Neither current nor life-course health can be identified as a source of cross-national variation in the DI recipiency rates, while variables describing the generosity of the DI system have strong explanatory power. This explanatory power is driven by the large differences in DI generosity across countries as described by the OECD indicators.

This leads to a threshold interpretation (Croda and Skinner 2009): Our broad set of health variables rank individuals well by health within each country. The thresholds, however, beyond which DI benefits are granted, are country-specific and have almost no relation to health. They are products of institutional characteristics such as minimum benefit levels and assessment requirements.

References

Aarts, L. J. M., Burkhauser, R. V., & de Jong, P. R. (Eds.). (1996). *Curing the Dutch disease. An international perspective on disability policy reform.* Avebury: Aldershot.

Banks, J., Kapteyn, A., Smith, J. P., & van Soest, A. (2004). *International comparisons of work disability.* RAND Working Paper, WP-155.

Börsch-Supan, A. (2005). Work disability and health. In A. Börsch-Supan et al. (Eds.), *Health, ageing, and retirement in Europe – First results from the survey of health, ageing and retirement in Europe* (pp. 253–258). Mannheim: MEA.

Börsch-Supan, A. (2008). Changes in health status and work disability. In A. Börsch-Supan et al. (Eds.), *Health, ageing, and retirement in Europe – Starting the longitudinal dimension* (pp. 228–236). Mannheim: MEA.

Croda, E., & Skinner, J. (2009). *Disability insurance and health in Europe and the US.* Paper prepared for the SHARE User Conference, October 2009, Mainz, Germany.

OECD. (2003). *Transforming disability into ability.* Paris: OECD.

Chapter 20
Health Insurance Coverage and Adverse Selection

Philippe Lambert, Sergio Perelman, Pierre Pestieau, and Jérôme Schoenmaeckers

20.1 Adverse Selection and Health Insurance

The term adverse selection is used in the insurance literature to describe a situation where an individual's demand for insurance (either the propensity to buy insurance, or the quantity purchased, or both) is positively correlated with the individual's risk of loss (e.g. higher risks buy more insurance), and the insurer is unable to allow for this correlation in the price of insurance. This may be because of private information known only to the individual, or because of regulations or social norms which prevent the insurer from using certain categories of known information to differentiate prices (e.g. the insurer may be prohibited from using information such as gender or ethnic origin or genetic tests). To test for the presence of adverse selection one checks the conjecture that contracts with more comprehensive coverage are chosen by agents with higher accident risk.

The problem is that such a test can as well reveal the presence of moral hazard. Like adverse selection moral hazard results from information asymmetry. In the case of adverse selection the informational issue concerns the individual's risk; in the case of moral hazard; it concerns the individual's behavior. Moral hazard occurs when the party with more information about its actions or intentions has a tendency or incentive to behave inappropriately from the perspective of the party with less information. *Ex ante*, it will be less cautious; *ex post*, it will seek overcompensation. In health insurance ex post moral hazard is likelier than *ex ante* moral hazard.

In the recent years, several authors have tested the relation between risk and insurance and have shown that the relation is not as clear as suspected. Chiappori and Salanié (1997, 2000) find no evidence of adverse selection in the automobile insurance market. Their main finding is that, although unobserved heterogeneity on risk is probably very important, there is no correlation between unobservable riskiness and contract choice. In other words, when choosing their automobile

P. Lambert (✉), S. Perelman, P. Pestieau, and J. Schoenmaeckers
CREPP – HEC Université de Liège, Bd. du Rectorat 7 (B31), 4000 Liege, Belgium
e-mail: p.lambert@ulg.ac.be; sergio.perelman@ulg.ac.be; p.pestieau@ulg.ac.be; jerome.schoenmaeckers@ulg.ac.be

insurance contracts, individuals behave as though they had no better knowledge of their risk than insurance companies. They are the first to show that the risk-coverage correlation can be either sign and to stress the need of a new model. Similarly, in the life insurance market, Cawley and Philipson (1999) do not find evidence of adverse selection.

On the opposite side, Finkelstein and Poterba (2004, 2006) find evidence of adverse selection in the UK annuity market. Along the same lines, Olivella and Vera Hernández (2006) observe the presence of adverse selection in British insurance markets, especially in private health insurance markets. Clearly the debate is wide open.

In this paper we intend to investigate on the basis of the information collected by SHARE if there is a relation, and if so what is its sign, between health risk and insurance coverage. In other words we are not planning to go beyond a simple statistical description of the relation between these two variables controlling for various characteristics of the concerned individuals.

For this purpose, SHARELIFE and wave 2 of SHARE contain valuable information. On the one hand, SHARELIFE surveyed retrospectively individuals about long periods of ill health or disability over their whole life. On the other hand, WAVE 2 questioned the same individuals on the characteristics of their health insurance coverage. Combining the answers given to these questions we estimate, for each country and for selected health care kinds, the relationship between health risks and insurance coverage, using a simple logistic model with full coverage as the dependent variable.

Two previous studies analyzed voluntary private health care insurance using SHARE data: (Paccagnella et al. 2008) and (Bíró 2010). They however have a different concern. Paccagnella et al. (2008) analyze the effect of having a voluntary health insurance policy on out-of-pocket spending for individuals aged 50 or more. They show that private insurance policy holders do not have lower out-of-pocket spending than the rest of the population. They also find that the main determinants of private insurance purchase are different in each country and this reflects the differences in the underlying health care systems. Bíró (2010) is interested by the presence of moral hazard.

20.2 Health Insurance Coverage

In wave 2, individuals were asked the following question: *Who finally pays for health care: yourself only, mostly yourself, mostly your health insurance, or your health insurance only?* This question was repeated for several kinds of care: medical visits to doctors (general practitioners or specialists), hospitalization (in public or private hospitals), nursing care (at home or in nursing homes), as well as for dental care and prescription drug expenditures. In this study we are particularly interested in full health insurance coverage (*paid by your health insurance only*) for the first three kinds of care indicated above.

20 Health Insurance Coverage and Adverse Selection

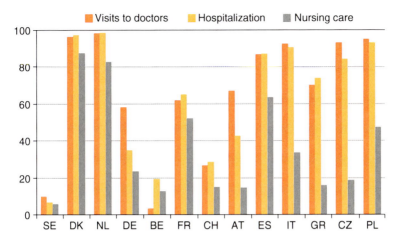

Fig. 20.1 Health care full insurance coverage (%)

Figure 20.1 reports, for each country, the percentage of individuals reporting full health insurance coverage for visits to doctors, hospitalization and nursing care. In each case, we consider an individual to be fully covered if he/she reports full coverage for at least one of the kind of care considered, e.g., hospitalization in a public or a private hospital. The analysis is limited to the 50–79 years old individuals who participated in both, wave 2 and SHARELIFE.

From Fig. 20.1 and depending on the kind of care, SHARE countries can be classified according to the proportion of the aged population with full health insurance coverage. For visits to doctors and hospitalization coverage, a first group includes the Denmark (DK), The Netherlands (NL), Spain (ES), Italy (IT), Poland (PL) and Czech Republic (CZ), which reach rates close to or higher than 90%; a second group is composed by Sweden (SE), Belgium (BE) and Switzerland (CH) with percentages lower than 30%; and finally, a third group comprises the rest of countries reporting intermediate rates of full coverage. For nursing care, the situation is dramatically different. In six countries – SE, BE, CH, AT, GR and CZ – less than 20% of respondents reported full coverage.

As expected, this classification is highly driven by national health care insurance institutions. As reported in Bíró (2010), all SHARE countries analyzed here have universal mandatory health insurance, private for Switzerland but public for the others, the only exceptions are the Netherlands (NL) and Spain (ES) where high earners are excluded from public health insurance. But universal coverage does not mean full coverage, in several cases cost sharing is the rule. Moreover, it happens that people can not take a private insurance to cover their participation in health costs, e.g. Belgium for visits to doctors, subscribing to a private insurance for health expenditures not covered by social protection schemes (co-payment) is forbidden; and in others this is allowed, e.g. Germany, where a fixed co-payment amount, 10 €, is charged in every quarter a doctor is visited (independent of total consumption).

Nevertheless, out of specific institutional regulations, in most cases full coverage is likely to be the result of a private individual decision. This is the main assumption we are making here. We postulate that individuals' health insurance behaviour is revealed by full coverage, compared with lower levels of coverage, and potentially affected by health risk expectations, as well as by other factors like gender, age, education and economic status. Another assumption is that respondents did not make systematic mistakes in reporting their health coverage status. Given the health insurance complexity, a potential measurement error bias exists which is probably related to individuals past experience with health providers and health insurance issues.

20.3 A Health Risk Indicator

SHARELIFE adds a valuable retrospective dimension to SHARE, particularly on health status. For the purpose at stake in this study, the availability of retrospective information opens the possibility to build innovative indicators of individuals' health risks based on their own health status experience over previous periods of their life. By construction, these indicators would likely be more suitable than those built on the basis of contemporaneous information exclusively. We argue here that individuals' perception of health risks, a latent variable, is correlated with past health experiences which potentially affected individuals' health insurance behaviour.

Our choice of this indicator relies on the number of long periods of ill health in adulthood. They are reported by SHARELIFE respondents who were invited to answer the following question: *Apart from any injuries you've already told us about today, as an adult, how many periods of ill health or disability have you had that lasted for more than a year: none, one, two, three or more, have been ill or with disabilities for all or most of my life?*

From the original answers to this question, we computed the percentage of individuals who suffered one or two or more long term spells of ill health (the remaining category corresponds to individuals with no spells of ill health). Figure 20.2 reports the average country percentages. The two extreme cases are, on the one hand, Switzerland (CH), with the lowest share of individuals with spells of ill health and, on the other hand, the Czech Republic (CZ), with nearly one third of the 50–79 cohort reporting at least one long term spell of ill health over their life span. Note that the variability of this indicator is high within countries across age and educational categories (not reported here).

20.4 The Health Risk Insurance Coverage Correlation

In order to test the potential correlation between health risks and health insurance coverage we proceed with the estimation of logistic models with full insurance coverage as the dependent variable. For each country, we estimate separately three

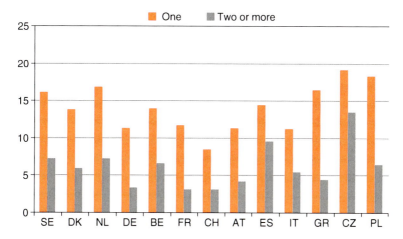

Fig. 20.2 Long-term periods of ill health (%)

Table 20.1 Logistic model: health risk variable parameter

Country	Visits to doctor Parameter	(P-value)	Hospitalization Parameter	(P-value)	Nursing care Parameter	(P-value)
AU	0.132	(0.229)	0.189	(0.047)**	0.106	(0.378)
BE	0.072	(0.551)	−0.137	(0.056)*	0.128	(0.050)**
CH	−0.101	(0.462)	−0.109	(0.422)	0.059	(0.715)
CZ	−0.047	(0.635)	−0.001	(0.988)	−0.201	(0.011)**
DE	0.059	(0.456)	−0.039	(0.636)	0.066	(0.459)
DK	0.310	(0.165)	0.139	(0.521)	−0.147	(0.051)*
ES	0.317	(0.011)**	0.313	(0.014)**	−0.094	(0.117)
FR	0.231	(0.011)**	0.255	(0.008)***	0.246	(0.013)**
GR	−0.130	(0.030)**	0.014	(0.836)	−0.498	(0.001)***
IT	0.161	(0.230)	0.545	(0.002)***	0.096	(0.123)
NL	−0.198	(0.266)	−0.416	(0.014)**	−0.097	(0.239)
PL	0.240	(0.205)	0.223	(0.168)	−0.016	(0.823)
SE	0.141	(0.127)	−0.042	(0.779)	0.187	(0.148)

***, **, *: Significant at the 1%, 5% and 10% level, respectively

models of full insurance coverage: visits to doctors, hospitalization and nursing care, respectively.

Table 20.1 reports the estimated parameters associated to the selected health risk indicator, which is the number of long-term ill health and disability periods, with values ranging from 0 (*none*) to 4 (*have been ill or with disabilities for all or most of my life*). The estimated parameters must be interpreted as marginal effects of increasing health risk on full health insurance coverage. A positive value would confirm the prevailing of individuals' adverse selection behaviour, and negative or statistically non significant parameters the absence of this relation. Several control variables were added to the model: age, gender, education and wealth quartiles.

With the exception of France, no other country exhibits positive and significant parameters for the three health insurance branches analyzed here. However, we found also positive and significant parameters in the case of Spain (ES) for visits to doctors, in the case of Austria (AU), Spain (ES) and Italy (IT) for hospitalization and finally in the case of Belgium (BE) for nursing care. And these results appear to be driven mainly by health care institutions' regulations than by adverse selection behaviour. This is the case of France and Spain, for instance, where individuals with heavy diseases, like cancer, are fully reimbursed.

We proceed to some sensitivity analysis. First, similar results were obtained using alternative models in which the health risk indicator was represented by categories instead of a continuous variable. Second, also comparable results, but with even less significant parameters, were obtained with the sample restricted to the population aged 50–64 years old in order to limit the potential selectivity bias due to higher mortality rates among people in bad health. Third, separate regressions were performed by gender and the results generally confirmed those reported in Table 20.1, with the only exception of full hospitalization coverage in the case of men, for which near all the parameters were not significant. We also tried to explain our results using the main characteristics of national health care systems, which are regularly published by the OECD (2009). Nothing significant resulted from this exercise.

20.5 Conclusions

Summing up, it appears that with a few exceptions there is no evidence of a huge adverse selection problem in health insurance among European elderly people. And even the exceptions are likely driven by full public health coverage offered by some European countries for specific risks. Nevertheless, both adverse selection and moral hazard are a key issue to be taken into account in the design of health care insurance, mainly in the ongoing debate on long-term care insurance, public or private, in many countries. This has lead governments and insurance companies to offer lump-sum reimbursement and not full reimbursement in case of dependence. Such an arrangement is clearly unsatisfactory but can only be explained by adverse selection and *ex post* moral hazard, which are more pervasive in long term care than in acute health care.

Certainly this short paper offers very preliminary results that need to be confirmed when new waves of SHARE are available. In the meantime, what clearly appears from this exercise is that SHARELIFE and SHARE data combined offer a rich framework for future research on adverse selection in health insurance.

References

Bíró, A. (2010). *Voluntary private health insurance and health care utilization of people aged 50+*. Central European University, Budapest, mimeo.

Cawley, J., & Philipson, T. (1999). An empirical examination of information barriers to trade in insurance. *American Economic Review, 90*, 827–846.

Chiappori, P. A., & Salanié, B. (1997). Empirical contract theory: The case of insurance data. *European Economic Review, 41*, 943–950.

Chiappori, P. A., & Salanié, B. (2000). Testing or asymmetric information in insurance markets. *Journal of Political Economy, 108*, 56–79.

Finkelstein, A., & Poterba, J. (2004). Adverse selection in insurance markets: Policyholder evidence from the UK annuity market. *Journal of Political Economy, 112*, 183–208.

Finkelstein, A., & Poterba, J. (2006). *Testing for adverse selection with unused observables*. National Bureau of Economic Research WP 12112.

OECD. (2009). *Health at a glance 2009 – OECD indicators*. Paris: OECD.

Olivella, P., & Vera Hernández, M. (2006). *Testing for adverse selection into private medical insurance*. The Institute for Fiscal Studies WP06/02.

Paccagnella, O., Rebba, V., & Weber, G. (2008). *Voluntary private health care insurance among the over fifties in Europe: A comparative analysis of SHARE data*. Department of Economics "Marco Fanno" WP 86, University of Padua.

Chapter 21
Lifetime History of Prevention in European Countries: The Case of Dental Check-Ups

Brigitte Santos-Eggimann, Sarah Cornaz, and Jacques Spagnoli

21.1 Access to Dental Care: An Indicator of Performance of Health Care Systems

European countries differ in various aspects of their health care systems, including the level of public investments and the importance given to prevention in national health policies. However, little is known regarding the association between the dissemination of preventive practices in populations included in the SHARELIFE project and public policies in health domain.

Access to care is a major characteristic for the performance of health care systems. Its assessment includes measurements of the reported use for a range of ambulatory and hospital services considered as an indicator of realized access. Some aspects of health care are particularly sensitive to the risk of inequality in access. It is the case of dental care. Indicators of access to dental care are part of recent WHO and OECD reports on health and care (WHO 2010; OECD 2009); they show a strong relationship with individuals' socio-economic circumstances (De Looper and Lafortune 2009; Van Doorslaer and Masseria 2004). Previous results from an analysis of SHARE baseline data indicated that, in participating European countries, visits to the dentist were unrelated to gender but increased with the level of education (Santos-Eggimann et al. 2005). Preservation of the mastication capacity is crucial to avoid denutrition and its consequences on health and function in older age. While in most countries European recommendations are issued for yearly dental check-ups, dental care remains costly, it is mainly offered by professionals active in the private sector and its coverage by social health insurances varies between countries from no one in Spain to complete in Austria (OECD 2009).

This chapter has the first objective to describe holes in the lifetime use of routine preventive dental controls in the SHARELIFE population, by country. The second aim is to confront differences in underuse of dental check-ups observed across

---

B. Santos-Eggimann, S. Cornaz, and J. Spagnoli
Health Services Unit; Institute of Social and Preventive Medicine, University of Lausanne, Route de Berne 52, 1010 Lausanne, Switzerland
e-mail: brigitte.santos-eggimann@chuv.ch; sarah.cornaz@chuv.ch; jacques.spagnoli@chuv.ch

European countries with contextual indicators that describe national human resources and reimbursement policy regarding dental care.

21.2 Data on Individuals and Health Systems

SHARELIFE, the third wave of SHARE data collection, included original retrospective questions on individual lifetime use of dental care. They concentrated on dental check-ups and care regularly performed from childhood to the time of interview. Information was collected on the age of respondents when starting to visit a dentist regularly, the average frequency of dental visits, periods of interruption and related age category of respondents, as well as reasons for not having dental check-ups and care regularly over the life course. For each respondent, we defined holes in the lifetime use of routine preventive dental controls as periods in which dentists were either not regularly visited, or visited less frequently than every 2 years. These periods were characterized by the age of respondents: during childhood (0–15 years), youth (16–25 years), young adulthood (26–40 years), adulthood (41–55 years), middle age (56–65 years) and older age (66–80 years).

Analyses of SHARELIFE individual data were performed on respondents aged 52–80 years from 13 countries participating in SHARELIFE, taking account the complete design information available from SHARELIFE and Share wave 2 to compute a correct estimate of variance. In order to take account of cohort effects, analyses were conducted for respondents aged 52–65 years and 66–80 years separately.

Results are displayed in four geographic regions: Northern (Sweden SE, Denmark DK, Netherlands NL), Continental (Belgium BE, Germany DE, Austria AT, France FR, Switzerland CH), Southern (Spain ES, Italy IT, Greece GR) and Eastern (Czech Republic CZ, Poland PL) countries.

A total of 21,281 respondents born between 1928 and 1956 were included in analyses. The sample size ranged from 711 in AT to 2,286 in GR. Respondents were categorized into two age classes, 52–65 years (57.0%) and 66–80 years (43.0%). The proportion of women was 55.7% in the first and 52.9% in the second age category.

Context variables describe national manpower (density of dentists: number of dentists in practice/1,000 inhabitants) from 1960 to 2006, according to 2008 OECD Health data, and the coverage of dental care (out-of-pocket dental expenditure as a percentage of total dental expenditure) in 2006, according to OECD indicators (OECD 2009). Additional information on health policy for selected countries was collected from WHO European observatory of health systems reports (WHO 2010).

21.3 Lifetime Underuse of Dental Care in European Countries

Both the level and the pattern of the lifetime underuse of dental care of middle aged individuals, as measured from self-reported information, varies widely between European regions. Figure 21.1 shows that, over the lifetime, underuse of dental care

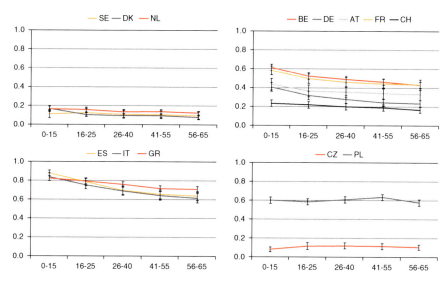

Fig. 21.1 Proportions of SHARELIFE respondents aged 52–65 years in 2008 who underused dental care at given ages

in subjects currently aged 52–65 years was at the lowest level and stable in all three Northern countries. Along the whole life course, a slightly higher (although not significantly) proportion of underuse is reported in NL as compared to DK and SE, except in the childhood period. In all three countries, underuse does not exceed 17% at all ages, with very close proportions observed in DK and SE.

Countries of the Continental region of Europe share a common pattern of lifetime experience: BE, DE, AT, FR and CH all experience a decline in underuse of dental care over the life course. However, the level of underuse varies substantially between these countries. It is the highest in BE and FR, where proportions decline from 62% in childhood to 43% at age 56–65 years, with a slightly higher level in BE than in FR. Underuse decreases during the life course from 44% in childhood to 33% at age 56–65 years in AT, and from 41 to 23% in DE. It is the lowest and characterized by the smallest decline over the life course in CH (from 23 to 17%). Figure 21.1 also shows that the ranking of countries in Continental Europe is maintained all over the life course, with Belgian respondents experiencing the highest level of underuse from their childhood to their middle age, and Swiss respondents reporting the lowest level of underuse during their whole life.

Like in Continental Europe, countries in the Southern region display a decline in underuse of dental care with age. However, their level is very high, with proportions of underuse ranging between 82 and 88% in childhood, and between 62 and 71% at age 56–65 years. GR experience a particularly high proportion of underuse of dental care in adulthood, while ES and IT share a similar experience of underuse along the whole life course.

In Eastern Europe, CZ and PL display a similar pattern of underuse unaffected by age. The level is low (between 8 and 10%) in CZ, where it compares with the experience of Northern countries. By contrast, it is high in PL (between 58 and 63%), where it compares with Southern countries in adulthood.

The life course experience of dental care in European populations seems to have improved over time in all countries participating in SHARE. A comparison between Figs. 21.1 and 21.2 suggests a cohort effect: in all countries except SE, the level of underuse reported by individuals aged 66–80 years (Fig. 21.2) is higher during the whole life course, than the level reported by the 52–65 years old (Fig. 21.1).

In the Northern region, differences between countries are larger than in the younger age category, and Fig. 21.2 displays an age effect that was not observed in Fig. 21.1: underuse of dental care reported by individuals aged 66–80 years was higher in their childhood and youth than in their adulthood and in older age. NL shows the highest level of underuse, ranging between 45% in childhood and around 30% from age 16–25 to age 76–80 years, and SE the lowest level (from 19 to 10%).

In the Continental region, the ranking of countries is similar in Figs. 21.1 and 21.2 except for DE. Underuse of dental care declines from 79% in childhood to 59% at age 76–80 years in BE, and from 48 to 34% in CH. However, the lifetime evolution of underuse for 66–80 year old individuals in DE is characterized by a high proportion of underuse in childhood (59%, rank 3/5) to a low proportion in the adulthood and older age (25% at age 76–80 years, rank 5/5).

A high level of underuse over the life course is registered in Southern countries for respondents aged 66–80. The order of countries is the same as the order observed for the younger age category of respondents; differences between countries, however, are larger and increasing over the life course. The level of underuse declines slightly from 94% in childhood to 85% at age 76–80 years in GR. The slope is steeper in ES (from 94 to 76%) and in IT (from 90 to 68%).

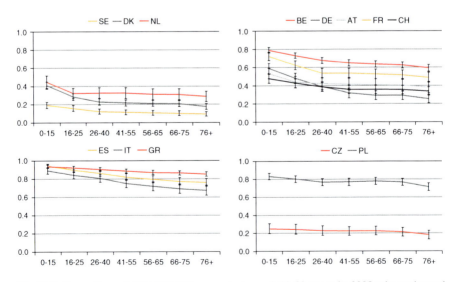

Fig. 21.2 Proportions of SHARELIFE respondents aged 66–80 years in 2008 who underused dental care at given ages

In Eastern Europe, the life course decline of underuse reported by individuals aged 66–80 is moderate, and the level of this indicator is much lower in CZ (from 25% in childhood to 18% at age 76–80 years) than in PL (from 83 to 72%).

21.4 Elements of Health Policy Regarding Dental Care in European Countries

In most countries participating in SHARELIFE, there is no clear evidence of a relationship between the level of underuse of dental check-ups and care and the density of dentists as shown in Fig. 21.3. While the level of underuse reported by respondents aged 52–65 is very close over their whole life course in the three Northern countries, a sizable difference in dentists' density is observed between NL (low density) and SE (high density), with DK in an intermediate position. This ranking, however, reflects the order of countries observed in the older age group, with a lower level of underuse over the life course in SE, a higher level in NL, and an intermediate level in DK. Differences today are smaller, with a level at 0.5 dentists/1,000 inhabitants in NL and 0.8 in SE and DK. In these Northern countries, the current proportion of dental expenditure paid out-of-pocket is estimated at 63% in SE and at 69% in DK (no data available in NL).

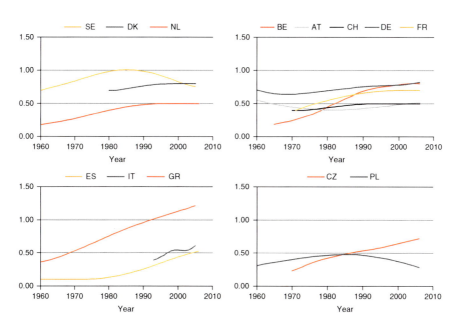

Fig. 21.3 Number of dentists in practice/1,000 inhabitants) from 1960 to 2006 or last year available, by country (source: OECD Health data 2008)

Countries in the Continental region have less difference in dentists' density. DE has the highest density over the whole 1960–2006 period. BE was characterized by a low density in 1965–1969 and a continuous increase in the following 40 years; it ranks as high as DE in the most recent year (0.8). FR also increased its density in the past 30 years and ranks slightly below DE and BE in 2005 (0.7). AT and CH are in the lowest range among these countries, with a quite stable density that compares to the density of NL today (0.5). Despite their increasing dentists' density over time, and despite a proportion of dental expenditure paid out-of-pocket in the low range (BE 34% and FR 28%), these two countries have the highest level of underuse of dental care in the Continental region. By contrast, DE, cumulating a high dentists' density over the last 45 years and the lowest out-of-pocket share of dental care expenditure, has a moderate level of underuse. The Swiss situation, where dentists' density is in the middle range and dental care expenditures are only marginally financed by the community (out-of-pocket share 91%), is particular, with a favorable position regarding underuse of dental care.

Among Southern countries, ES was characterized by a particularly low density of dentists from 1960 to the middle of the years 1980. Density then increased up to 0.5 in 2006. This level compares with IT in the recent years (no data available before 1993). By contrast, the dentists' density in GR was at 0.4 in 1960 and increased to 1.2 in 2005. The very high proportion of the population experiencing underuse of dental care along the whole life course in ES may be explained not only by a low density of professionals, but also by the fact that dental care is almost exclusively paid out-of-pocket (97%) in this country. Information regarding the share of dental care expenditure paid out of pocket in GR and IT is not available in the OECD database.

In the Eastern region, the evolution of dentists' density differs between CZ, where it increased from a low 0.2 dentists/1,000 inhabitants in 1970 to 0.7 in 2006, and PL, where dentists' density increased from 0.3 in 1960 to 0.5 in 1977–1991, and then declined to reach 0.3 in the end of the 1990 years. The share of out-of-pocket financing of dental care is lower in CZ (30%) than in PL (69%), which may explain the much higher level of underuse of dental care over the life course in PL.

21.5 Reasons for Missing Routine Dental Controls in European Countries

SHARE participants reporting holes in their lifetime use of routine preventive dental controls were also asked about causes. In all countries, a majority of respondents mentioned that they did not perceive a need, it was not usual to have dental check-ups, or information was lacking (Fig. 21.4).

The proportion mentioning that they had no place to receive this type to care near home did not exceed 5%, except in GR (18%) and in PL (12%). Overall, financial

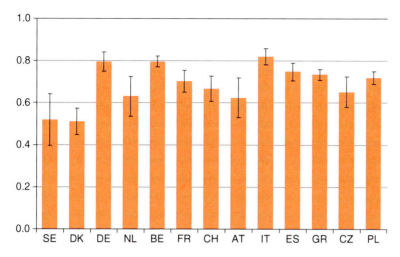

Fig. 21.4 Proportion of respondents reporting holes in their lifetime use of routine preventive dental controls who mentioned that that they did not perceive a need, it was not usual to have dental check-ups, or information was lacking

reasons (cost of care or lack of insurance coverage) were mentioned in 16%, but the proportion exceeded 20% in CH (30%), ES (29%), SE (28%), DK (24%) and GR (21%).

21.6 Conclusions

Underuse of dental care over the life course compromises health in old age and should be limited to improve the quality of life. SHARELIFE data indicate that even in countries characterized by the most favorable conditions of professional manpower and care reimbursement, and using a very conservative estimate of underuse of dental care (periodicity of controls at least every 2 years was the cut-off used in this analysis, while most countries recommend yearly visits to the dentist), underuse of dental care concerns at least 10% of the population aged 52–80 years.

Large variations were observed between European regions and countries, with levels of underuse reaching more than 85% along the whole life course in some countries. In most cases, high levels of underuse can be explained by a dentists' density lower than 0.5/1,000 inhabitants and / or by a large proportion of dental care bills paid out-of-pocket. Intriguing situations, like CH where the level of underuse is moderate and the proportion of dental care expenditure financed out-of-pocket is very high, may have roots in a sufficient dentists' density conjugated with a high level of income. In GR, the high level of underuse of dental care despite a very high density of dentists may be related to the high cost of this type of care (Siskou et al. 2009) or to variations in density by region (e.g. Athens 1.66/1,000 inhabitants, Ionian Islands 0.57/1,000 inhabitants in 2001, according to data provided by the Hellenic Statistical Authority (EL.STAT.)). Further investigations should be dedicated to gain a better understanding of the specific situation in countries or regions, and to integrate the potential multiple causes of underuse of dental care in Europe.

References

De Looper, M., Lafortune, G. *Measuring disparities in health status and in access and use of health care in OECD countries.* OECD health working papers 43 (March 9, 2009).
Hellenic Statistical Authority (EL.STAT.), http://www.statistics.gr (accessed April 2010).
OECD Health data (2008). Statistics and indicators for 30 countries. Jun 26, 2008. OECD: Paris.
OECD Health at a glance (2009) – OECD Indicators. Dec 8, 2009. OECD: Paris.
Santos-Eggimann, B., Junod, J., & Cornaz, S. (2005). Health services utilization in older Europeans. In A. Börsch-Supan et al. (Eds.), Health, ageing, and retirement in Europe – First results from the survey of health, ageing and retirement in Europe (pp. 133–140). Mannheim: MEA.
Siskou, O., Kaitelidou, D., Economou, C., Kostagiolas, P., & Liaropoulos, L. (2009). Private expenditure and the role of private health insurance in Greece: Status quo and future trends. *European Journal of Health Economics, 10,* 467–474.
Van Doorslaer, E., & Masseria, C. Income-related inequality in the use of medical care in 21 OECD countries. OECD health working papers 14 (May 11, 2004). Paris.
WHO European Observatory on Health Systems and Policies http://www.euro.who.int/observatory (accessed March 4, 2010).

Chapter 22
Disparities in Regular Health Care Utilisation in Europe

Nicolas Sirven and Zeynep Or

22.1 Regular Care Use as a Public Health Issue

A standard doctor visit or routine check-up can be essential for maintaining good health. People who have regular checkups may identify health issues well before any symptoms show up and receive the treatment for reducing onset and complications. Many of the costly and disabling conditions can be prevented through early detection. Therefore, improving access to routine checkups is considered as an objective for the health care systems (WHO 2002). For instance, women are advised to have regular gynaecological visits from an early age and mammography from 50 onwards. Major guidelines recommend periodic comprehensive evaluation of blood values and regular follow-up of blood pressure after a certain age (Mandel et al. 2000). Persons who have regular eye examinations may experience slower decline in vision and functional status (Gohdes et al. 2005).

Despite common recommendations and quasi universal health care coverage in all European countries, there are large differences in the utilization patterns of different health services. Even across countries with similar levels of GDP (Gross Domestic Product) per capita, the rate of using recommended services varies significantly. Moreover, there is a large body of work showing that in many countries the probabilities of seeing a doctor (and the number of visits) are not identically distributed across socio-economic groups after correcting for differences in the need for care. More specifically, a "pro-rich" bias in the use of specialist care is well demonstrated in Europe (Van Doorslaer et al. 2004, 2006). At the same time, studies show that the magnitude and direction of these inequalities vary significantly from one country to another (Hanratty et al. 2007). This may reflect different strategies for setting up and coordinating preventive and curative health care services. The level of available health resources and their organisation varies significantly across welfare

---

N. Sirven (✉) and Z. Or
Institute for Research and Information in Health Economics, IRDES, 10, rue Vauvenargues, 75018 Paris, France
e-mail: sirven@irdes.fr; or@irdes.fr

states. However, the link between the organisation of health care resources and their long-term utilisation is not well understood.

In particular, little comparative information is available on different types of health service utilisation such as routine health check-ups and variations in utilisation patterns over a longer time span to compare the shift (if any) in healthcare habits of different generations. First, most studies examine the variations in care utilisation at one point in time since the usual datasets do not allow for analysing respondents' long term health care habits. Second, very few studies proposed a cross-country analysis of the disparities in different types of care utilisation. It remains unclear whether variations in health service utilisation are a generalized phenomenon, or whether these inequalities are observed only for some services, countries and demographic groups. Third, there is little information on the evolution of health care utilisation habits of different generations and the role of health care policies in determining these utilisation patterns.

SHARELIFE provides a unique source of internationally comparable information on individual's long-term health care utilisation patterns as well as their life-course social, economic, and health status. It is important to identify the patterns of regular care use within and between countries and analyse the determinants of use in order to improve health policies in welfare states. The objective of this study is twofold: first, it aims to compare and investigate the determinants of health care utilisation habits over the life span of individuals across European countries. Second, the study proposes an analysis of the impact of country-specific time related macroeconomic factors which characterize welfare states. In particular, we test the role of health sector development with respect to general economic growth in determining healthcare utilisation habits.

22.2 Measuring Healthcare Utilisation Habits

SHARELIFE provides some original information on the health care consumption habits of individuals over their life course. In particular, respondents are asked whether or not they had regular health check-ups over the course of several years. For instance: "Have you ever had your blood pressure checked regularly over the course of several years?" This differs from the usual questions on health care consumption asking if the respondents had consumed health care over a specific period (usually over the past year). The respondents are asked if they have regular check-ups for six types of care. Santos-Eggimann et al. in Chap. 21 address the issue of dental check-ups over the life course, while we focused on the five other types of care: blood pressure, blood tests, vision tests, and (for women only) gynaecological visits and mammograms. Our dependent variables are binary taking the value 1 if the respondent ever had regular health check-ups and 0 otherwise. In the descriptive analysis, the variable "age when regular health care started" is also taken into account; but this variable is not used as dependant variable in this chapter. Additional variables retained in the multivariate analysis are described below.

22 Disparities in Regular Health Care Utilisation in Europe   243

The information provided in SHARELIFE allows comparing the consumption patterns of different generations. In order to examine the change in healthcare utilisation patterns, we have constructed three cohorts observing the age distribution of respondents in our sample: Cohort 1 consists of people who were born between 1925 and 1934, Cohort 2 consists of those who were born between 1935 and 1944 and Cohort 3 corresponds to those born after 1945.

This data is complemented by the information collected in SHARE wave 1 (2004–2005) and 2 (2006–2007) providing data on the current life circumstances of individuals aged 50 and over in 15 European countries. The sample used in this study was restricted to respondents (1) who were interviewed in at least one of the first two waves, and re-interviewed in SHARELIFE; and (2) whose understanding of questions asked in SHARELIFE was satisfying (442 cases were deleted, 1.6% of the initial sample). The final sample includes 22,251 respondents (96% of the initial sample) from 13 countries covering four Euro-regions: North (Sweden, Denmark, the Netherlands), East (Czech Republic, Poland), Continental (Belgium, Germany, Austria, France, Switzerland), and South (Spain, Italy, Greece) of which 12,128 women.

At the individual level, we have information on both the initial and current life circumstances which might have a direct or indirect impact on individuals' care consumption habits. Moreover, we can control for general respondents characteristics: gender, age, having children and current and lifetime health status as natural determinants of care utilisation. The health status is assessed by the following variables:

- Current self-rated health (SRH): a dummy taking the value 1 if the respondents perceive her health as excellent or very good at wave 3, and 0 otherwise
- Chronic conditions: a dummy taking the value 1 if the respondent reports 2 or more chronic illnesses (cancer, diabetes, etc.) at wave 1 or wave 2, and 0 otherwise
- SRH at 10 years old: self-rated retrospective value of health, taking the value 1 if the respondent reports that health during childhood was in general excellent or very good, and 0 else (i.e. good, fair, or poor, or spontaneously "Health varied a great deal")
- Periods of ill health or Ever physically injured: a binary index of health, taking the value 1 if the respondent reports any periods of ill health over the life-cycle (>1 year) or if she reports any physical injury over the life-cycle (>1 year)
- For vision tests only: (1) whether or not the respondents wear glasses, (2) a dummy taking the value 1 if the respondent declares her eyesight for seeing things at a distance (like recognising a friend across the street) is excellent or very good, and 0 otherwise.

In order to capture the socio-economic conditions of the individuals we used the following:

- Labour market situation: (1) A dummy indicating if the respondent ever worked, and (2) a dummy taking the value 1 if the respondent is still at work at wave 3, and 0 otherwise

- A comfort index made out of 6 items (whether or not the household's accommodation had the following when the respondent was 10 years old: fixed bath, cold running water supply, hot running water supply, inside toilet, central heating, and whether or not there was a room by person) taking theoretical values between 0 (none of them) and 6 (all of the items)
- A dummy indicating if the respondent encountered any periods of Financial Hardship throughout her life;
- Assets at interview time (cross sections of three waves): the average amount of assets in Euros the respondent reports over the first two waves of SHARE. We use assets instead of last-year income, since this variable is a better indicator of economic well-being of individuals over the life course. Total assets have a smoother evolution over the life course and they discriminate better than yearly incomes which become less informative after a certain age (due to retirement and common pension schemes);
- Education: highest level of education completed (in three categories: none or primary, secondary, and tertiary);
- A set of country by cohort dummies were included in the models.

At the country level, we are interested in the role of economic development versus healthcare system in determining healthcare utilisation habits. Four variables were considered at the country level: (1) GDP per capita, (2) Total health expenditure per capita, (3) Public expenditures on health, and (4) the density of practicing physicians. Country-specific time series are constructed using several editions of the OECD Health database covering the period 1975–2005. For the purposes of the regression analysis, each series is divided into three sub-periods corresponding to the economic development and health care provision for three cohorts of individuals aged in their 50s. Thus, for Cohort 1, we measure economic growth and health care supply for the period 1975–1985, for Cohort 2 the period is 1985–1995 and for Cohort 3 it is 1995–2005. For each indicator we calculated (1) the average volume/level over 10 years, (2) the mean average annual growth rate over 10 years, and (3) Total growth rate over 10 years.

22.3 Regular Health Care Use at a Glance

Significant differences in regular health care utilisation are observed across countries and gender (Table 22.1) and across different age groups. Figure 22.1 compares the share of population having regular check-ups by age, in four country groups. Northern countries (Sweden, Denmark and the Netherlands) have lower rates of regular health check-ups for all indicators except for mammography for which the rates are significantly higher than all other countries. Southern countries (Spain, Italy, Greece) followed by the continental Europe have systematically higher check-up rates for blood tests, blood pressure tests, vision tests and gynaecological visits. The prevalence of regular health care check-ups seems to increase with age in the case of blood

22 Disparities in Regular Health Care Utilisation in Europe

Table 22.1 Population having ever had regular health check-ups, by gender, in %

Country	Blood tests Women	Blood tests Men	Blood pressure tests Women	Blood pressure tests Men	Vision tests Women	Vision tests Men	Gynaeco. tests Women only	Mammograms Women only
Austria	72.0	68.7	63.2	61.7	77.9	64.9	74.9	66.1
Germany	69.4	69.9	65.1	61.3	71.4	66.8	78.9	45.9
Sweden	41.4	52.7	54.6	61.7	50.5	44.3	81.7	88.9
Netherlands	51.8	53.0	62.7	61.6	63.4	57.1	47.9	83.1
Spain	85.4	82.8	81.9	73.7	70.0	64.7	59.8	68.5
Italy	80.8	79.0	75.7	75.8	59.3	52.4	57.3	57.7
France	78.9	77.0	87.3	88.5	84.8	82.3	71.9	75.2
Denmark	50.7	51.9	47.4	51.2	47.8	38.0	56.8	32.8
Greece	89.5	85.3	82.7	80.8	74.6	67.2	69.4	46.4
Switzerland	60.9	65.2	69.6	65.8	69.8	64.0	75.2	48.4
Belgium	83.3	84.4	84.6	84.6	76.7	71.7	69.9	71.6
Czechia	48.5	53.9	60.1	63.3	67.4	57.3	86.0	62.3
Poland	58.0	52.4	68.7	58.2	56.0	44.6	51.5	38.4
Total	72.3	71.6	73.1	70.8	68.8	63	67.1	58.6

Note: Calibrated individual weights used

tests and blood pressure tests. This could be explained by the decline in health status by age. These tests become more frequent as health status deteriorates. But, no cohort or age effect is found in the case of vision tests, which is surprising as often vision deteriorates after 50 years old. As expected, the prevalence of regular gynaecological visits and mammograms is higher for younger cohorts. It is interesting to note that the rate of regular mammogram use for the first cohort (oldest generation) in Northern countries is even higher than for the second cohort in all other countries which suggests that this specific preventive policy have been effectively adopted in these countries since the middle of the last century.

Additional information on the health care utilisation habits provides useful insights. Figure 22.2 demonstrates the shift in health care utilisation behaviour for three cohorts. It shows that (1) the mean age for starting regular health check-ups is decreasing at each new cohort (except for mammography), and (2) the prevalence/ use of regular health check-ups increases at each new cohort. For example, the age of starting regular check-ups for blood pressure has been dropped from after 70 years old for the first cohort (born between 1925 and 1935) to around 50 for the third cohort (born between 1945 and 1955). It is also interesting to note that for the later cohorts there is a little "peak" around 20 years old concerning blood tests, blood pressure and vision tests, suggesting that new cohorts (especially post-war ones) may have benefited from prevention policies at an early age. Regular gynaecological visits and mammograms follow a somehow different pattern since the period of start for these tests is age-specific: around 20 years old for the former (child bearing age) and around 50 years old for the latter. Therefore, no significant shift in starting age was expected. Nonetheless, there is a visible upward shift in the prevalence of women having regular gynaecological visits and mammograms at each new cohort. Such differences suggest a significant change over the past

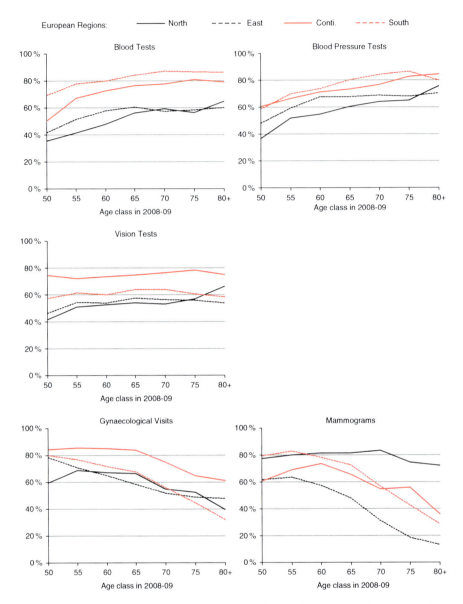

Fig. 22.1 Population having regular health check-ups, frequencies by Euro-regions and age-class
Source: SHARELIFE (2008–2009). Calibrated individual weights used

40 years in health care consumption habits of European populations which might partly explain improving health outcomes.

Figure 22.3 provides some information on the reasons given for not having regular health care check-ups. "Not considered to be necessary" is the main motive

22 Disparities in Regular Health Care Utilisation in Europe 247

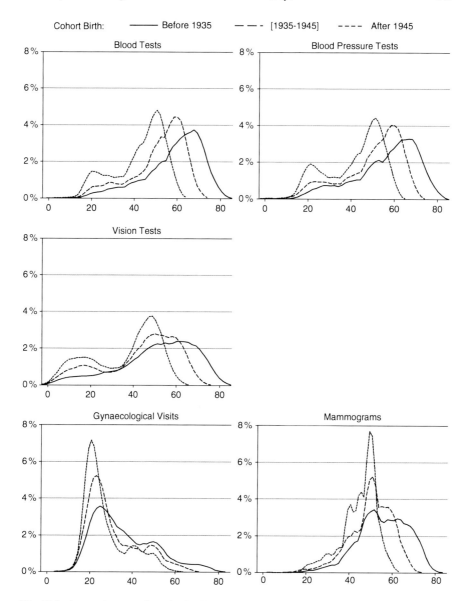

Fig. 22.2 Age at the start of regular health check-ups
Source: SHARELIFE (2008–2009). Calibrated individual weights used

cited to explain why respondents do not use regular health care: more than 80% of the cases for blood tests, blood pressure, and vision tests, and about 70% for gynaecological visits and mammograms. While this pattern remains constant across cohorts, there seems to be some variations across euro-regions in particular for

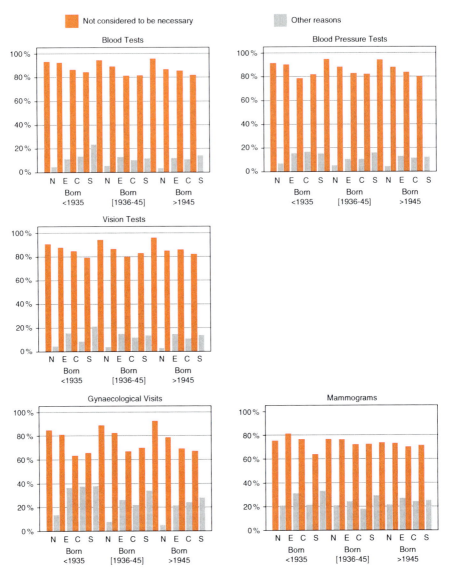

Fig. 22.3 Reasons given for non-regular health check-ups, frequencies by Euro-regions and cohort of birth
Source: SHARELIFE (2008–2009). Calibrated individual weights used

gynaecological visits. The results show that about 20–30% of the population have other reasons for not using regular care: not affordable, not covered by health insurance, did not have health insurance, time constraints, not enough information about this type of care, not usual to get this type of care, no place to receive this type of care close to home, etc. Clearly, the importance of these issues depend directly on the health care system design and need to be tackled by appropriate health policies in different countries. Furthermore, the item "Not considered to be necessary" could also capture confounding reasons like being in good health, or having little information about prevention. Note that while the prevalence of regular health check-ups has been increasing at each generation (Fig. 22.2), the major reason for not having regular health care has not changed over the three generations. This suggests that there is room for improvement in all countries through public information and education strategies.

22.4 Determinants of Individual Healthcare Habits

A general finding in the literature is that privileged people in terms of socio-economic conditions (education, income, etc.) have a higher propensity to use specialist care. Although this result is well established on cross-sectional and panel data (where care utilisation is investigated over the last year or the last 6 months), little is known on variations in different types of health services which can have a direct impact on individuals' health and wellbeing. Moreover, health care utilisation over the life-cycle may have a different pattern than health care consumption at a given point. SHARELIFE retrospective data allow for examining consumption habits of individuals over their life course. In order to establish the determinants of regular health care utilisation at the individual level, separate Logit models are run for each dependant variable indicating whether or not individuals ever had regular blood test, blood pressure, vision test, gynaecological visits and mammography. The models control for the following variables at the individual level: general individual characteristics (having children, age and gender), health status, socio-economic conditions (Box 2). Moreover, a series of country and cohort dummies are used for taking into account unobserved heterogeneity across countries and cohorts. Note that in the present analysis individual level observations are nested naturally in cohorts and in countries. Hence we define 39 clusters (C) corresponding to the interaction between countries ($J = 13$) and cohorts ($T = 3$).

Results of the logistic regression analysis (Table 22.2) suggest that, all else being equal, men have higher propensity to have regular blood pressure tests than women but lower propensity to have regular vision tests. In addition, controlling for the cohort effects, gynaecological visits and mammograms decrease with age (within cohorts). Age is not significant for other regular health care check-ups. We also note that women with children have significantly higher propensity to have regular gynaecological visits and mammograms, which may suggest that having children has a longer term impact on women's care utilisation habits.

The results concerning the impact of health are consistent with the literature: reporting an excellent or very good health status at wave 3 is associated with a lower propensity to have regular blood pressure tests and blood tests, while no impact on the other types of services. On the other hand, having two or more chronic illnesses increases the odds of using all types of health services regularly including vision tests. Moreover, having experienced long periods of ill health or

Table 22.2 Determinants of regular health care use (odds-ratios, logit estimates)

Dependent variable	Blood pressure	Blood tests	Vision tests	Gynec. visits	Mammo-grams
Male	1.076**	1.106	0.744***		
Age at wave 3	1.030*	1.026	1.016	0.954**	0.942*
With children	1.116	1.134	1.121**	1.487***	1.226***
Secondary education	1.005	1.004	1.281***	1.373***	1.357***
Tertiary education	0.962	1.040	1.401***	1.544***	1.592***
SRH when child	1.153***	1.149***	0.992	0.983	1.024**
Periods of ill health injured	1.396***	1.441***	1.210***	1.095***	1.127***
SRH at wave3	0.796***	0.817***	0.987	1.054	1.028
2+ chronic illnesses	2.521***	2.597***	1.365***	1.111**	1.260***
Wear glasses	–	–	2.544***	–	–
Eyes distance (excellent/ v.good)	–	–	0.877***	–	–
Did you ever work? (ref. = yes)	1.224***	1.432***	1.246***	1.383***	1.605***
At work at wave 3	1.007	0.951***	1.187***	1.099	0.874**
Periods of financial hardship	0.932	0.984	1.006	1.048*	0.995
Childhood comfort index	0.995	0.989	1.011***	1.063***	1.005
Assets quartile 2	1.015	1.032	1.188***	1.111***	1.106***
Assets quartile 3	1.062	1.097***	1.158***	1.218***	1.422***
Assets quartile 4	0.974*	1.084***	1.149***	1.263***	1.489***
Obs.	22,251	22,251	22,235	12,128	12,128
Pseudo R²	0.112	0.137	0.092	0.112	0.190

Note: *** $p < 1\%$; ** $p < 5\%$; *$p < 10\%$. Country-cohort fixed effects included but not shown in the table. Categories not shown are reference categories (female, without children, primary education, and assets quartile 1

having been severely injured appear to increase significantly the propensity to use regular health care. Note that the retrospective self-rated health status in childhood is also associated with regular blood pressure and blood test use, as well as regular mammograms: good child health increases the propensity to have these tests regularly.

As to the impact of socio-economic conditions, we first note that, all else being equal, the impact of socio-economic variables is stronger for vision tests, gynaecological visits, and mammograms which are performed by specialists or depend on referral from specialists. Blood pressure tests, usually carried out regularly by generalists, appear to be distributed more equitably. Second, people with high levels of assets have significantly higher propensity to use regularly all of the health services, except blood pressure tests. Controlling for other socio-economic variables and health status, the odds of having regular gynaecological visits are 26% higher for people with highest level of assets (fourth quartile) compared to those with lowest asset levels. Furthermore, controlling for assets levels, higher levels of economic comfort during childhood also seem to increase the odds of having regular vision tests and gynaecological visits. Third, even after the impact of economic conditions taken into account, the education appears to be a significant determinant of regular care utilisation. All else being equal, the odds of having regular gynaecological visits and mammograms are 50% higher for women having tertiary education compared to those having only primary education. Finally, having a job or being in the labour market has a mixed effect on regular health care use. On the one hand, having ever worked is the most important determinant of regular care utilisation for all services. The odds of having regular blood test are 43% higher for people who have had a job one time in their life compared with those who have never worked. The Odds ratios are 1.6 for mammography, 1.4 for gynaecological visits and 1.2 for blood pressure and vision tests. This may reflect the existence of preventive policies introduced through work place regulations but also the insurance status which may depend directly on work status in some countries. On the other hand, controlling for age, having a job at the time of the survey seems to reduce propensity to have regular blood test and mammography. This may suggest the higher time cost of health care for those who are actively in the labour market.

22.5 Exploring Cross-Country Differences in Healthcare Habits

The results from Table 22.2 show that, adjusted for the individual differences in health and living conditions, there is still significant heterogeneity in regular care utilisation between countries and cohorts. For example, the Logit coefficients presented in Fig. 22.4 give the propensity of using regular blood tests by country and by cohort, after controlling for individual characteristics of respondents. They indicate that there is a north–south gradient in the propensity to have regular blood tests. These findings corroborate previous findings from Fig. 22.1, and allow disentangling country effects from cohort effects.

In order to explain these differences in regular care use across countries and cohorts, we estimated panel data models with time fixed effects. The coefficients of country/cohort clusters (fixed effects) for each type of regular heath care are used as a new dependant variable to be regressed on a set of country-cohort level context variables. Cohort 1 for Poland and the Czech Republic were removed from this analysis because some context variables were not available for the period 1975–1985 for these two clusters.

In these models, we test for the impact of general economic development of a country (GDP) and the resources devoted to health care system on health care utilisation patterns of country/cohorts. As presented before we have three variables for measuring overall health system resources. These health care variables are introduced one by one in the equations together with the GDP. This probably captures better the general economic and health care conditions for each cohort in different countries. It also allows "isolating" the effect of health care policies on regular care use from the confounding influence of economic growth. For sake of simplicity, only the significant results are displayed in Table 22.3 (Comprehensive results are available from the authors upon request).

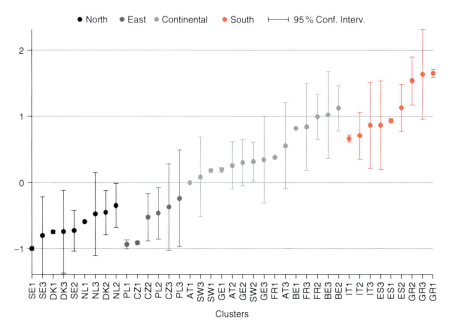

Fig. 22.4 Propensity to have regular blood tests, by cohort and country
Note: Fixed effects controlled for a set of individual variables. Reference is AT1 (Austria, Cohort 1: Before 1935)

The results suggest that physician density has a significant impact on the utilisation of most health services. The propensity to have regular blood pressure, blood tests, vision tests, gynaecological visits is significantly higher in country/cohorts where the number of physicians per capita is higher. However, concerning mammography, there is no significant impact of physician density on their regular utilisation. This is not surprising since in most countries, breast cancer screening is also carried out within specific targeted programmes mobilising different resources, while all the other services require a physician contact/visit. In addition, controlling for GDP growth, individuals who were in countries and cohorts where the average growth rate in health expenditure was higher, have a higher propensity to use regular health check-ups for blood pressure, blood and vision tests. The health expenditure growth (in real terms) reflects probably the overall investment effort in healthcare by period and by country and may indicate improvement in availability (easier access) of services. It is interesting to see that controlling for health care resources, GDP growth does not have any significant impact on individuals' care consumption habits. There is even a small negative impact on the use of blood pressure tests, which may suggest that during the periods of rapid economic growth, time cost for healthcare is higher and less attention is paid to health (Ruhm 1996). When time invariant effects are taken into account (cohort effects), cross-country differences in Europe in terms of the prevalence of regular health care utilisation is partly explained by national strategies regarding the provision of practising physicians and overall investment in health care.

Table 22.3 Determinants of cross-country differences in regular health care use (odds ratios)

Dep. var: coeffs. of clusters from logit models	Blood pressure	Blood tests	Vision tests	Gynec. visits	Mammograms
Model 1					
GDP per capita, mean average annual growth rate (over 10 years)	0.810*	0.84	0.84	0.842	0.993
THE per capita, mean average annual growth rate (over 10 years)	1.265**	1.244*	1.210**	1.038	1.069
Model 2					
GDP per capita, total growth rate (10 years)	0.982	0.985	0.986	0.986	0.997
THE per capita, total growth rate (10 years)	1.017**	1.016*	1.014**	1.003	1.007
Model 3					
GDP per capita, average (over 10 years)	0.999***	0.999***	0.999***	1.000	1.000
Practising physicians – density/1,000 pop. average (over 10 years)	1.400***	1.525***	1.241*	1.331*	1.053
Time fixed effects model					
Model 4					
GDP per capita, mean average annual growth rate (over 10 years)	0.784*	0.817	0.807	0.844	1.037
THE per capita, mean average annual growth rate (over 10 years)	1.291***	1.263*	1.230**	1.021	1.059
Model 5					
GDP per capita, total growth rate (10 years)	0.979*	0.982	0.982	0.986	1
THE per capita, total growth rate (10 years)	1.019**	1.018*	1.015**	1.002	1.006
Model 6					
GDP per capita, average (over 10 years)	0.999***	0.999***	0.999***	1.000	1.000
Practising physicians – density/1,000 pop. average (over 10 years)	1.962***	2.160***	1.289	1.719**	1.282

Note: THE = Total health expenditure. GDP and THE at NCU 2000 GDP price levels.
***$p < 1\%$; **$p < 5\%$; *$p < 10\%$

22.6 Conclusion

This study provides some new evidence on the variations of health care utilisation habits of different cohorts in 13 European countries. We found that while there is a general shift toward more regular and preventive care utilisation in all countries; there are still significant differences between countries and cohorts.

Our results confirm that there are significant social inequalities in the life time regular utilisation of health care services allowing for early detection and prevention, after correcting for differences in the need for care over the life-cycle. Individuals with higher levels of education and assets have a higher propensity to have regular use of blood tests, vision tests, gynaecological visits, and mammograms. The impact of education is significant even after controlling for income and occupation. We also find that social inequalities are stronger for services provided by specialists.

There is also evidence that, once the individual effects have been isolated, cross-cohort and country differences in the prevalence of regular care use are partly associated with national health policies. Controlling for GDP growth, physician density also appears to be a significant determinant of regular utilisation of all health services except for mammography. Moreover, countries and cohorts which have experienced higher growth rates of total health expenditures have higher prevalence of regular blood pressure tests and regular vision tests. In contrast, the impact of overall economic growth on health care utilisation habits appears to be insignificant if not negative.

These results suggest that there is significant room for public health policies for reducing disparities in regular use of health services within and across European countries. Health promotion and education can play an essential role for assuring equal and timely treatment of diseases within and across countries. Moreover, strengthening primary care provision appears to be critical for improving health systems' ability to provide and develop services in a timely manner.

References

E Doorslaer, Van, Masseria, C., Koolman, X., & OECD Health Equity Research Group. (2006). Inequalities in access to medical care by income in developed countries. *Canadian Medical Journal, 174*(2), 177–183. doi:10.1503/cmaj.050584.

Gohdes, D. M., Balamurugan, A., Larsen, B. A., & Maylahn, C. (2005). Age-related eye diseases: An emerging challenge for public health professionals. *Preventing Chronic Disease, 2*(3), A17.

Hanratty, B., Zhang, T., & Whitehead, M. (2007). How close have universal health systems come to achieving equity in use of curative services? A systematic review". *International Journal of Health Services, 37*(1), 89–109.

Mandel, J. S., Church, T. R., Bond, J. H., Ederer, F., Geisser, M. S., Mongin, S. J., et al. (2000). The effect of fecal occult-blood screening on the incidence of colorectal cancer. *The New England Journal of Medicine, 343*(22), 1603–1607.

Ruhm, C. J. (1996). *Are recessions good for your health?* National Bureau of Economic Research (NBER), Working Paper, no.5570.

Van Doorslaer, E., Masseria, C., & the OECD Health Equity Research Group Members (2004). *Income-related inequality in the use of medical care in 21 OECD countries.* OECD Health Working Paper, 14.

WHO (2002). Integrating prevention into health care. Fact Sheet. No. 172. Geneva.

Chapter 23
Does Poor Childhood Health Explain Increased Health Care Utilisation and Payments in Middle and Old Age?

Karine Moschetti, Karine Lamiraud, Owen O'Donnell, and Alberto Holly

23.1 The Long-Lasting Health Care Consequences of Childhood Conditions

There is growing evidence of a large, positive and significant association between health experienced in childhood and that evolving in adulthood. For example, Case et al. (2005), using data from a British cohort followed from birth, find that poor health in childhood is correlated with reduced health in adulthood up to the age of 42. The association remains after controlling for socioeconomic circumstances in both childhood and adulthood. One of the many potential implications of such a correlation is that the health care costs of an ageing population will, in part, be determined by health events experienced in childhood and may be less responsive to the prevention and treatment of health problems that arise in adulthood. On a more positive note, the health care costs of ageing may be lower than anticipated since cohorts that will reach old age in the coming years have experienced better childhood health and health care than did their predecessors. This paper examines directly the extent to which health care utilisation and payments in middle and old age are predictable from childhood health experiences.

The analysis is made possible by the rich data provided by SHARELIFE, which provides detailed retrospective life histories, including health events in childhood

K. Moschetti (✉) and A. Holly
Institute of Health Economics & Management (IEMS), University of Lausanne, Bâtiment Vidy, Route de Chavannes 31, CH-1015 Lausanne, Switzerland
e-mail: Karine.Moschetti@unil.ch; Alberto.Holly@unil.ch

K. Lamiraud
Department of Economics, ESSEC Business School, Avenue Bernard Hirsch, B.P. 50105, Cergy 95021, France
e-mail: lamiraud@essec.fr

O. O'Donnell
Department of Balkan, Slavic and Oriental Studies, University of Macedonia, Egnatia 156, 54006 Thessaloniki, Greece
e-mail: ood@uom.gr

that can be linked to the contemporaneous SHARE panel data on health, health care utilisation and payments for health care for populations aged 50+. The cross-national nature of SHARE makes it possible to examine whether the ability of childhood health to predict health care utilisation at older ages varies across European countries. This would be expected if cross-country variation in the universality and quality of health care during the childhood of the SHARE cohorts was sufficiently marked such that childhood health problems were more likely to be effectively treated within some systems than in others. The correlation between health in childhood and adulthood may also vary across countries if there was variation in the extent to which poor childhood health disrupted education. A curtailed education may have long term health consequences through health knowledge, behaviour and economic circumstances.

Examination of cross-country variation is an indirect way of investigating the extent to which the long-lasting health care consequences of childhood illness can be mitigated. A more direct approach is to test whether constrained access to health care in childhood is correlated with reduced health and increased health care use in later life. The SHARELIFE data are somewhat limited in the information provided on health care use in childhood, but we use that which is available to test the hypothesis.

Section 23.2 describes the data drawn from SHARE and SHARELIFE. Section 23.3 presents the association between childhood health status and health care use and payments in middle and old age, while Sect. 23.4 investigates the possible mechanisms responsible for this correlation. Section 23.5 examines whether the association between childhood health status and health care use varies across groups of European countries. Section 23.6 discusses the relevance of our findings for policy and future research.

23.2 Data

We use health care utilisation and payment data from the first and second waves of the SHARE survey combined with retrospective data on childhood health from SHARELIFE. The analysis is based on a pooled sample of respondents for which information on all required variables is reported. The sample consists of 25,737 individuals, of whom 13,438 were interviewed in both waves of SHARE, 9,528 were interviewed in wave 1 only and 2,771 were interviewed only in wave 2. All were interviewed in SHARELIFE. Overall, the sample comprises 39,175 observations from 13 countries (Austria, Germany, Sweden, Netherlands, Spain, Italy, France, Denmark, Greece, Switzerland, Belgium, Czech Republic and Poland). The proportion of women is 56% and the mean age is 63.

We use the following dichotomous indicators of health in childhood (defined as age less than 16): (1) whether the respondent reports his/her health status during childhood as "fair" or "poor" versus "good", "very good", or "excellent"; (2) whether he/she reports to have had at least one of the following childhood illnesses – polio, asthma, respiratory problems other than asthma, severe diarrhoea, meningitis/encephalitis, chronic ear problems, speech impairment and problems with vision (question HS008); (3) whether he/she reports to have had at least one childhood illness or condition from another list – severe headaches or migraines; epilepsy, fits or seizures; emotional, nervous or psychiatric problems; broken bones, fractures; appendicitis; childhood diabetes or high blood sugar; heart trouble; leukaemia or lymphoma; cancer or malignant tumour (question HS009); (4) a dichotomous variable indicating whether he/she had at least one inpatient stay of 1 month or longer, or had three or more hospital stays over a period of 12 months, during childhood. Furthermore, we use whether the respondent's parents smoked during his/her childhood as an indicator of exposure to a health risk. Finally, access to health care during childhood is measured by whether or not the respondent reports that s/he did not have a usual source of care during childhood. Admittedly, this is a rather crude indicator of constrained access to care, but, as will be demonstrated, it is nonetheless informative in explaining health and health care use in adulthood.

Health care utilisation (HCU) in middle and old age is measured by the reported number of contacts with a physician (either a General Practitioner or a specialist) during the previous 12 months. Out-of-pocket (OOP) payments include non-refundable expenses for inpatient care, outpatient care, prescribed drugs and nursing homes and monetary values are expressed in Euros, adjusted for purchasing power parity. This is more comprehensive than our measure of utilisation, in the sense that it covers more types of medical treatment. However, it will vary not only with the quantity of care but also with the prices paid, which vary with insurance coverage that, in turn, is dependent on country, age, health condition and economic circumstances. Apart from their correlation with utilisation, OOP payments are of interest because of the burden they place on the household. We seek to establish whether individuals afflicted by illness in childhood are carrying an economic burden of this, in terms of increased payments for health care, in middle and old age.

Health status in middle and old age is measured by standard indicators: (1) self-assessed health status collapsed into a binary indicator distinguishing good–excellent health from poor–fair health; (2) whether the respondent suffers from a diagnosed chronic illness; (3) whether he/she reports at least two symptoms (see note to Table 23.1 for list) (4) the number of limitations in activities of daily living and instrumental activities of daily living. In addition, we control for smoking status in adulthood through indicators of currently smoking and of having stopped smoking.

Measures of socioeconomic status in adulthood are education level, collapsed into three categories [Low level (International Classification of Education – ISCED – 1–2), Intermediate level (ISCED 3–4), High level (ISCED 5–6)], log annual gross household income per capita and employment status represented by an indicator of whether the individual is not working and is below the age of 65. Moreover, all specifications and models estimated include country dummy variables, as well as gender–age group dummies using five age groups: [45–54], [55–64], [65–74], [75–84], [85+].

Table 23.1 Association of number of doctor visits with childhood health, without and with control for adult health and socioeconomic status. Negative binomial regression on pooled wave 1 and 2 data

Dependent variable: Number of doctor visits	No control for adult health or SES	With control for adult health	With control for adult health and SES
Childhood health			
Health reported as fair or poor	0.1873***	0.0665***	0.0631**
Any illness from question HS009	0.0598***	0.0096	0.0097
Any illness from question HS008	0.0719***	0.0162	0.0184
Inpatient stay ≥ 1 month or ≥ 3 times in any year	0.0880***	0.0570**	0.0569**
Childhood health risks and health care			
Parent smoked	0.0372**	0.0273**	0.0267*
No usual source of health care	0.0748**	0.0027	0.0035
Adult health			
Health reported as good, very good or excellent		−0.439***	−0.429***
Smokes currently		−0.0899***	−0.0907***
Has stopped smoking		0.0347***	0.0393***
Has a diagnosed chronic illness		0.4460***	0.441***
Has at least two symptoms		0.337***	0.333***
Number of limitations in (instrumental) activities of daily living		0.0667***	0.0651***
Adult socioeconomic status			
Log household income per capita			0.0196***
Intermediate level of education (ISCED 3–4)			−0.0248
High level of education (ISCED 5–6)			−0.0240
Not working and less than 65			0.1401***
Observations	39,175	39,175	39,175
Chi-square test for the joint significance of childhood health and health risks variables, p-value	0.0000	0.0010	0.0015

Notes: *$p < 0.1$ **$p < 0.05$; ***$p < 0.01$ based on robust standard errors adjusted for clustering at individual level. All models included gender–age and country dummies. Childhood health in comparison to good, very good or excellent. Minor illnesses HS008 are excluded. Health in comparison to fair or poor. Symptoms form SHARE question PH010

23.3 Association Between Childhood Health and Health Care Utilisation in Adulthood

We begin by describing the total association between health status in childhood and doctor visits and private OOP payments for health care in middle and old age. Figure 23.1 plots the mean number of doctor visits by age for those reporting good–excellent health (brown) and those reporting poor–fair health (blue) during childhood. The profiles are strikingly different. Around the age of 50, those who report poor–fair health during childhood consult the doctor at least 60% more than

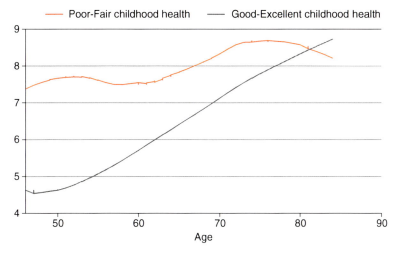

Fig. 23.1 Mean number of doctor visits by age split by self reported health status before age 16

those who report good–excellent health in childhood. Those reporting good–excellent health in childhood are above the age of 70 before they reach the utilisation rate of those reporting poor–fair health at age 50. There is a clear upward trend in doctor visits with age for the sample reporting good–excellent health in childhood. This is much less apparent for the group reporting poor–fair health. Consequently, by age 80 there is no difference between the groups in their use of doctors.

In early middle age individuals who report poor health in childhood appear to be already in a very poor state of health and are making intensive use of health care. They may already have contracted the chronic conditions that will materialise only in old age for individuals who were healthy in childhood. If so, this points to very large differences in inherent health, which are already evident in childhood. However, it could also be that causality runs in the opposite direction. It may be that individuals experiencing severe health problems in middle age and making frequent use of health care are more likely to remember childhood health problems and to report poor health in childhood.

Similar patterns are observed for the age profiles of doctor visits when childhood health is measured by other indicators, such as having had a childhood illness, condition or inpatient stay. We can examine the association between each indicator and health care utilization and payments using regression analysis.

Firstly, we run a negative binomial model of the number of doctor visits (Cameron and Trivedi 1986) on the pooled data set including our complete set of indicators of childhood health and circumstances, while also controlling for gender–age group dummies and country dummies. Given that we have pooled wave 1 and 2 data, standard errors are computed allowing for clustering at the level of the individual. The results (Table 23.1, column 1) confirm a significant and strong positive correlation of childhood ill-health with the number of doctor visits in middle and old age. All four indicators of health status in childhood are individually strongly significant. In addition, variables proxying health risks and access to health care during childhood are associated with more visits to the doctor. That is, having had a parent who smoked and not having had a usual source of health care during childhood are both associated with increased visits at the 5% level of significance.

Secondly, we estimate a two-part model of OOP expenditures (Duan et al. 1983; Jones 2000) consisting of a Logit model to explain the probability of having a positive payments (Table 23.2) and a least-squares regression for positive (log transformed) OOP amounts (Table 23.3).

Indicators of childhood health status are highly jointly significant in explaining both the probability of incurring OOP payments and their level. Having experienced at least one illness from both of the lists of childhood illnesses is significantly correlated with a higher probability of paying OOP for health care (Table 23.2, column 1). Childhood

Table 23.2 Association of propensity to incur any out of pocket expenses for health care with childhood health, without and with control for adult health and socioeconomic status (logit model)

Dependent variable: 1 if any OOP for health care	No control for adult health or SES	With control for adult health	With control for adult health and SES
Childhood health			
Health reported as fair or poor	0.0778	−0.0277	−0.0181
Any illness from question HS009	0.1635***	0.1135***	0.1003***
Any illness from question HS008	0.2306***	0.1587***	0.1345***
Inpatient stay ≥1 month or ≥3 times in any year	0.0046	−0.029	−0.0230
Childhood health risks and health care			
Parent smoked	−0.0324	−0.0388	−0.0385
No usual source of health care	−0.1076*	−0.1431**	−0.1220*
Adult health			
Health reported as good, very good or excellent		−0.1352***	−0.1763***
Smokes currently		−0.2690***	−0.2668***
Has stopped smoking		0.0161	0.0094
Has a diagnosed chronic illness		0.5875***	0.5866***
Has at least two symptoms		0.5433***	0.5538***
Number of limitations in (instrumental) activities of daily living		−0.0388**	−0.0326*
Adult socioeconomic status			
Log household income per capita			0.0946***
Intermediate level of education(ISCED 3–4)			0.1668***
High level of education (ISCED 5–6)			0.2346***
Not working and less than 65			−0.0334
Observations	39,175	39,175	39,175
Chi-square test for the joint significance of childhood health and health risks variables, p-value	0.0000	0.0001	0.0014

Notes: *$p < 0.1$ **$p < 0.05$; ***$p < 0.01$ based on robust standard errors adjusted for clustering at individual level, All models included gender–age and country dummies. Childhood health in comparison to good, very good or excellent. Minor illnesses HS008 are excluded. Health in comparison to fair or poor. Symptoms form SHARE question PH010

Table 23.3 Association of positive amount of OOP payments with childhood health, without and with control for adult health and socioeconomic status (OLS regression)

Dependent variable: Log of positive OOP payments	No control for adult health or SES	With control for adult health	With control for adult health and SES
Childhood health			
Health reported as fair or poor	0.1450***	0.0237	0.0276
Any illness from question HS009	0.0762***	0.0389*	0.0310
Any illness from question HS008c	0.0480*	0.0064	−0.0075
Inpatient stay ≥1 month or ≥3 times in any year	0.0690*	0.0444	0.0498
Childhood health risks and health care			
Parent smoked	0.0417**	0.0326*	0.0314*
No usual source of health care	0.0570	0.0020	0.0137
Adult health			
Health reported as good, very good or excellent		−0.3260***	−0.3490***
Smokes currently		−0.0227	−0.0218
Has stopped smoking		0.0140	0.0116
Has a diagnosed chronic illness		0.3741***	0.3712***
Has at least two symptoms		0.2479***	0.2543***
Limitations in (instrumental) activities of daily living		0.1292***	0.1329***
Adult socioeconomic status			
Log household income per capita			0.0686***
Intermediate level of education (ISCED 3–4)			0.0988***
High level of education (ISCED 5–6)			0.1705***
Not working and less than 65			0.0566**
Observations	27,611	27,611	27,611
Chi-square test for the joint significance of childhood health and health risks variables, p-values	0.0000	0.0198	0.0408

Notes: *$p < 0.1$ **$p < 0.05$; ***$p < 0.01$ based on robust standard errors adjusted for clustering at individual level. All models included gender–age and country dummies. Childhood health in comparison to good, very good or excellent. Minor illnesses HS008 are excluded. Health in comparison to fair or poor. Symptoms form SHARE question PH010.

illness is also associated with a higher level of OOP payments, as is having experienced a long, or repeated, stay in hospital during childhood (Table 23.3, column 1).

Conditionally on having positive OOP expenses, having had a parent who smoked is associated with a higher level of OOP payments, at the 5% level (Table 23.3, column 1). There is no evidence that lack of access to a usual source of health care in childhood is associated greater payments for health care in adulthood. In fact, this variable is negatively associated with the probability of paying out of pocket for health care.

The three regressions confirm a significant association between health and health risks in childhood and health care utilization and payments in middle and old age. Individuals who report poorer childhood health status, childhood illness, extended or repeated inpatient stay during childhood, or being exposed to parental smoking visit their physicians more often and incur higher OOP payments for health care beyond the age of 50. The regressions also provide some evidence that health care in childhood is associated with health care utilisation in middle and old age. Those that did not have a usual source of health care during childhood visit the doctor more often as adults.

23.4 Exploring Mechanisms Responsible for the Association

Why is childhood health status correlated with health care utilisation and payments in middle and old age? The most obvious mechanism is through adult health, which, according to the existing literature, could be direct (Kuh and Wadsworth 1993; Barker 1995) and/or via socioeconomic status (SES) (Marmot et al. 2001; Case et al. 2005). Another possible mechanism is from childhood health problems to socioeconomic status and subsequently health care seeking behaviour, conditional on health care needs. In spite of the universal nature of European health care systems, the widespread commitment to equity in access to health care and the relatively low level of OOP payments in many countries, socioeconomic differences in the utilisation of health care, particularly specialist care, remain (Van Doorslaer et al. 2004). These inequalities may reflect, amongst other factors, educational differences in health expectations and knowledge of health care. But conditional on measured adult health and SES, health care utilisation may remain correlated with childhood health. In part, this could arise from a direct causal effect, perhaps because an early-life experience of ill-health and health care permanently influences preferences for the receipt of care from doctors and hospitals. Of greater importance, most likely, is an ability of reported child health to provide information on variation in health status over and above that which can be captured by the indicators included in a survey, even one so rich in health indicators as SHARE. In this case, childhood health indicators provide valuable information on inherent health status that could prove useful in a number of contexts. For example, it might be used as a determinant of the demand for health insurance.

With the aim of assessing the relative importance of these different mechanisms, we observe how the association between the childhood health indicators and doctor visits and OOP payments change as we sequentially add controls for adult health status (column 2 in Tables 23.1–23.3) and then socioeconomic status (column 3 in Tables 23.1–23.3). If childhood health has a direct impact on health care use in middle and old age, or if it provides information on inherent health status over and above that contained in the indicators of adult health, then after the controls are added the childhood health variables will continue to have significant effects on health care use at older age.

As expected, the indicators of adult health status are strong predictors of doctor visits (Table 23.1, column 2). Controlling for health in adulthood, the coefficients on the indicators of childhood health fall greatly in magnitude but, together with exposure to parental smoking, they remain jointly significantly correlated with doctor visits. Individually, poor–fair self-assessed childhood health status and long, or repeated, inpatient stays remain significant positive predictors of the number of doctor visits. Exposure to parental smoking also remains significantly correlated with increased visits to the doctor. Since the adult health indicators include current and past smoking status, this is suggestive of a direct impact of parental smoking on health care utilisation in adulthood over and above that through the influence on the persons own smoking behaviour. Lack of access to a usual source of health care during childhood is no longer significantly correlated with doctor visits once control is made for adult health status. This implies that lack of access to care in childhood is negatively correlated with health in adulthood and this is shown explicitly for all indicators of adult health in Table 23.4 (column 1). This is suggestive of inadequate treatment of childhood health problems having long terms consequences for health and, in turn, utilization of health care.

Adding controls for SES has less effect on the magnitude and the significance of the coefficients of the childhood health variables. The indicators remain highly jointly significant and both reported poor–fair childhood health status and childhood inpatient stay continue to be positively correlated with increased doctor visits, as does parental smoking at the 10% level of significance. These results suggest that childhood experience of illness has a long-lasting direct impact on health care seeking behaviour and/or that the childhood health indicators operate as proxies for current inherent health that is not fully captured by contemporaneous health indicators.

Table 23.4 Access to health care in childhood and its correlation with indicators of health status in adulthood by major European regions and average share of the population with total medical coverage in 1960

	All countries	Southern Europe	Central Europe	Northern Europe	Eastern Europe
Percentage with no usual source of health care during childhood	5.69%	8.37%	4.87%	3.23%	7.24%
Pearson's Correlation b/w not having usual source of health care in childhood and indicators of health in adulthood					
Health reported as good, very good or excellent	−0.0372*	0.0011	−0.0381*	−0.0013	−0.115*
Has a diagnosed chronic illness	0.0173*	−0.0017	0.0319*	0.0004	0.0826*
Has at least two symptoms	0.0285*	0.0073	0.0187*	0.0232*	0.1148*
Limitations in (instrumental) activities of daily living	0.0571*	0.0753*	0.0281*	0.0051	0.1196*
Observations	39,175	11,386	14,914	9,505	3,370
Average share of the population with total medical coverage in 1960 (OECD 2005)		61.70%	74.24%	88.70%	Not available

Notes: All models included gender–age and country dummies. Childhood health in comparison to good, very good or excellent. Minor illnesses HS008 are excluded. Health in comparison to fair or poor. Symptoms form SHARE question PH010
*Indicates Chi-square test rejects null of independence at the 5% level or lower

With respect to the explanation of the propensity to spend OOP on health care (Table 23.2, column 2), upon the inclusion of adult health status variables, most of the coefficients of the childhood health variables fall in magnitude but both indicators of illness in childhood remain significant and the set of childhood health measures are still highly jointly significant. Joint significance of the child health and health risks indicators is also maintained in explanation of the level of OOP payments, but in this case only two of the indicators remain individually significant at the reduced level of 10% (Table 23.3, column 2). As with doctor visits, the introduction of SES indicators has little effect on the childhood health coefficients and their joint significance.

In summary, we find that poor childhood health does predict greater health care utilisation and payments in middle and old age. This predictive ability falls substantially, but remains significant, when adult health status is controlled for. Controlling for education and socioeconomic status has a much smaller impact on the correlation, suggesting there is little impact from childhood health to SES and subsequently to health care utilisation. There is also some evidence that inadequate treatment of childhood health problems has long terms consequences for health and, in turn, utilization of health care.

23.5 Does the Association Vary Across Europe?

As suggested in the introduction, differences in the coverage and effectiveness of health care systems during the childhood of the SHARE cohorts may result in cross-country variation in the degree to which childhood health problems predict health care use in middle and old age. To investigate this, we distinguish four geographic areas: Northern (SE, DK, NL), Central (BE, DE, AT, FR, CH), Southern (ES, IT, GR) and Eastern (CZ, PL) countries. Figure 23.2 plots the age profile of mean number of doctor visits by self-assessed childhood health status for the four areas.

At all ages, the average number of visits to a doctor is lowest in the Northern European countries and is highest in Southern Europe. In all four geographic areas, those who report being in poor–fair health during childhood visit the doctor more often in middle and old age than those who report being in good–excellent health during childhood. The difference is much larger in Southern Europe than it is in the other regions, and while the disparity decreases as people get older in Northern, Central and Eastern Europe, this is not true in Southern Europe. These differences, at least in part, seem to be attributable to differences in the propensity to report poor childhood health status. Only 5% of respondents in the Southern countries report poor–fair health status in childhood whereas the respective percentages are 6.8%, 9.5% and 10% in Eastern, Northern and Central countries. Consequently, the disparity in health status between those reporting poor–fair and those reporting good–excellent would be expected to be greater in Southern countries and the greater disparity in health care utilization in this region may reflect this.

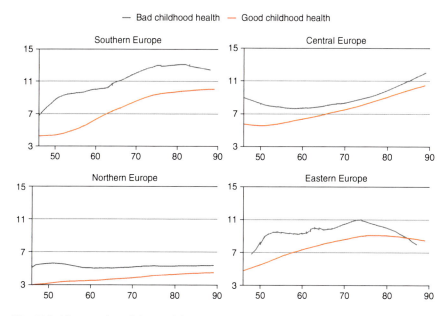

Fig. 23.2 Mean number of doctor visits by age split by health status in childhood for European regions

Running the same regression models for each geographic area separately reveals that, after controlling for adult health, childhood health and health risk indicators are significant predictors of doctor visits and OOP payments in Southern and Central Europe but not in Northern and Eastern Europe. Respondents from Southern and Eastern Europe are less likely to have had access to a usual source of care during childhood (Table 23.4, columns 2–5) and this is a predictor of higher OOP payments only in these two regions. Southern Europe is also the region in which, in 1960, the lowest proportion of the population was enjoying full medical coverage (Table 23.4). This suggests, although it certainly does not confirm, that the Southern European cohorts experienced less adequate treatment of childhood health problems resulting in long term consequences for health and, in turn, health care payments.

23.6 Conclusion and Policy Relevance

We have shown that childhood health is a strong predictor of health care utilization and payments in middle and old age. This predictive ability is considerably weakened when control is made for measured health in adulthood, although it remains significant. Controlling for socioeconomic status has little or no effect. The correlation is stronger in Southern and Central Europe than it is in Northern and Eastern countries.

The fact that correlation holds in Southern Europe, and in some extent in Central Europe, but not in Northern Europe may be interpreted as indicative of childhood health problems having longer lasting implications in countries that were further from universal health care during the childhood of the SHARE cohorts of older individuals. In 1960, when the youngest of the SHARE cohort was around 5 years old, the proportion of the population with total medical coverage was lowest in Southern Europe and highest in Northern Europe (Table 23.4). These cross European differences in coverage correspond both with those in the proportion of the SHARE-LIFE respondents reporting not having a usual source of health care in childhood and with the strength of the association between childhood ill-health and health care utilisation in adulthood. This suggests that the correlation between childhood health and adult health and health care utilisation may be weaker for cohorts younger than the SHARE sample. Medical coverage is now universal throughout Europe, the effectiveness of medicine has increased tremendously and barriers to access have been reduced. Childhood health problems would therefore be expected to leave less of a lasting impact than they did in the past.

Our analysis is pertinent to the much discussed health care costs of ageing populations. Ability to curtail these costs will depend, among other things, on the extent to which health care use in old age is driven by health experiences in childhood. Our results are consistent with childhood health conditions having long-lasting effects on health, and consequently utilisation of health care. If there are indeed such very long lived consequences of childhood illnesses, then there would be very long lag times in realizing the returns to timely and effective policy interventions. By the same token, the magnitude of these lifetime returns could be very large. Improving childhood health in populations now will lead to future cohorts costing less in old age than do their current counterparts.

But we should be careful about rushing to such bold implications. It could very well be that childhood health correlates with health care utilisation in old age because both reflect inherent health status. Variation in childhood health reflects the variation in the physiological and psychological robustness of individuals that is present throughout the lifetime and generates persistent differences in the utilisation of health care. On this interpretation, the correlation we observe does not necessarily indicate opportunities for long term returns to effective childhood health care. It does mean that variation in lifetime health care costs may be predictable from an early age. Depending upon whether information on childhood health problems is kept private, or must be shared with insurers, this will have important implications for the efficiency and/or equity of health insurance.

References

Barker, D. J. P. (1995). Fetal origins of coronary heart disease. *British Medical Journal, 311* (6998), 171–174.

Cameron, A., & Trivedi, P. (1986). Econometric models based on count data: Comparisons and applications of some estimators and tests. *Journal of Applied Econometrics, 1,* 29–54.

Case, A., Fertig, A., & Paxson, C. (2005). The lasting impact of childhood health and circumstance. *Journal of Health Economics, 24*, 365–389.

Duan, N., Manning, W. G., Morris, C. N., & Newhouse, J. P. (1983). A comparison of alternative models of the demand for health care. *Journal of Business and Economics Statistics, 1*, 115–126.

Jones, A. M. (2000). Health econometrics. In A. J. Culyer & J. P. Newhouse (Eds.), *Handbook of health economics* (Vol. 1, pp. 265–344). Amsterdam: Elsevier Science.

Kuh, D. J., & Wadsworth, M. E. (1993). Physical health status at 36 years in a British national birth cohort. *Social Science and Medicine, 37*(7), 905–916.

Marmot, M., Brunner, S., & Hemingway, S. (2001). Relative contributions of early life and adult socioeconomic factors to adult morbidity in the Whitehall II study. *Journal of Epidemiology and Community Health, 55*, 301–307.

OECD Health data (2008). Statistics and indicators for 30 countries. June 26, 2008. OECD: Paris.

Van Doorslaer, E., Koolman, X., & Jones, A. M. (2004). Explaining income-related inequalities in health care utilisation in Europe: A decomposition approach. *Health Economics, 13*(7), 629–647.

Part IV
Persecution

Chapter 24
Persecution in Central Europe and Its Consequences on the Lives of SHARE Respondents

Radim Bohacek and Michał Myck

24.1 Reflections of European History

The analysis of life history of individuals living in Europe would not be accurate and complete without considering the major historical events of the twentieth century. Many respondents in the SHARE sample have lived through periods of Nazi or Soviet occupation, direct World War II experience, the post-war period, and in the case of the Czech Republic, East Germany and Poland, through several decades of communism. In this chapter we provide the first results on effects these historical events have had on the life of individuals and their families in the SHARE sample.

To our knowledge, these issues have not been studied quantitatively in a general sample framework of the whole population. The existing literature mostly consists of historical research, case studies, and specialised samples of Holocaust survivors or patients suffering from post-traumatic stress disorder. The SHARELIFE data offer a unique and perhaps the last opportunity to analyse the effects of these important historical events on welfare of European populations.

24.2 Persecution in Europe in the Twentieth Century

Following Rummel (1994), we define persecution as "the responsibility of a government, regime, or self-governing group for an unarmed and non-physically threatening person's death, imprisonment, dispossession, deprivation of individual

---

R. Bohacek (✉)
The Economics Institute of the Academy of Sciences of the Czech Republic, Politickych veznu 7, 111 21 Prague 1, Czech Republic
e-mail: radim.bohacek@cerge-ei.cz

M. Myck
Centre for Economic Analysis, CenEA, ul. Krolowej Korony Polskiej 25, 70-486 Szczecin, Poland
e-mail: mmyck@cenea.org.pl

rights or freedoms". Because much of persecution occurred during wartime, we exclude combat deaths during war or military action, non-combatants that die as a by-product of military action, and punishment for what would normally be crimes.

One might perceive persecution as extreme cases of negative "welfare policies" of governments which are the subject of this publication. It is obvious that organised persecution has the most drastic and horrifying consequences on a person's life, physical and mental health, family, education, professional career and other aspects of their existence. These effects, apart from immediate short term consequences, may have important long-lasting implications. It is also well documented in Beebe (1975), Danieli (1998), and Krell et al. (1997) that these effects may bear consequences on younger generations. In this chapter we provide evidence on the scale, form and consequences of persecution as reported by the sample of SHARE-LIFE respondents. We provide an account of reports of persecution in all SHARE countries, but focus our analysis on three populations who were most likely to have suffered from persecution in the twentieth century – the Czech Republic, former East Germany and Poland.

We are aware of the difficulty of analyzing and comparing these acts of persecution across time, different countries and different regimes. Nevertheless, we believe that our quantitative analysis can provide important insights into the scale of the dramatic forms of state intervention into people's lives, and their consequences in terms of past and current wellbeing of the European populations.

24.3 European History and State Persecution in the Twentieth Century

In this section we provide a brief and incomplete list of types of state intervention in SHARE countries which would qualify as persecution according to the definition of Rummel (1994). They can be broadly related to World War II, the communist regimes, civil and colonial wars, and persecution in democratic regimes.

Among the first category belongs the persecution by the Nazi and Soviet regimes: the Holocaust, genocide and reprisals against occupied populations, imprisonment or concentration camps, forced labour and resettlement, forced military service, various forms of persecution affecting job prospects and education opportunities. All of these acts of persecution were also directed against these regimes' own population (also before the beginning of the war). The reasons for persecution were political, racial, religious, sexual orientation, class origin, or any other characteristics convenient for the regime in power. Among World War II we include also the post-war persecution of German nationals on the liberated territories either by the Soviet Army or the domestic majority population (murder, violence, forced resettlement and dispossession in the Czech Republic and Poland in 1944–1946).

Depending on the age of our respondents, the communist regimes in the Czech Republic (1948–1989), East Germany (1945–1989) and Poland (1944–1989) covered all or nearly all of childhood and a substantial proportion of adult lives of our respondents. The intensity of persecution varied greatly between regimes and time periods. The most intense periods of persecution occurred until 1956, with several hundred thousand persons affected by murder, labour camps, imprisonment, political trials, forced collectivization, resettlement and other acts of violence. In the later periods, the forms of persecution were less severe but continued in the form of restricted access to education or persecution at work, penitary military service, psychiatric confinement, and other restriction of civic freedoms (including travel). All or nearly all private property was confiscated by the government (except for Polish small scale farmers and shop owners). Relatively liberal periods (e.g. early 1968 in the Czech Republic) were followed by political clamp-down and periods of significant unrest and persecution all the way until the collapse of the communist rule in the late 1980s. The reasons for persecution were mostly political, religious and based on class origin.

The third category of persecution in SHARE countries is related to civil wars, military dictatorships and colonial wars. These include the Spanish civil war and its consequences, the Greek civil war, or the French, Dutch or Belgian loss of colonies, notably the French war in Algeria.

The last category covers the persecution in otherwise free and democratic societies in which people at different times experienced persecution because of sex, race, origin, religion, age or other reasons.

Importantly, there were several layers of society which were affected by persecution disproportionately than others: the Jewish population during the Holocaust and after (e.g. in 1968 in Poland), other religious and ethnic minorities, wealthy people, church members and intellectuals. Often the same individuals were persecuted first during World War II by the Nazis, and then by communist regimes after the war. The crucial aspect of persecution is that it was not only one member of the family that was affected but consequences were faced also by close relatives either directly or as a result of resettlement, job or educational restrictions.

24.4 State Persecution in the SHARELIFE Questionnaire

The current generation of SHARE respondents may include some of the last surviving individuals who have lived through World War II, the most oppressive years of communism and other instances of major turmoil. The SHARELIFE data may thus be one of the last opportunities to examine the consequences of these events on people's past and their current situation, in particular in an international context.

Significant numbers of individuals who have suffered persecution are no longer alive, many either as a direct consequence of oppression, or because persecution indirectly shortened their lives through effects on health and living standards. Moreover, persecution in almost all of its forms experienced in twentieth century Europe was far from "random", and usually wealthy or well educated families were more likely to suffer from it. Both of these factors would imply that any observed effects of persecution which we identify in the data would represent a lower bound of the overall consequences of persecution. If some individual characteristics positively affect the outcomes we measure and on the other hand were positively correlated with persecution the coefficients on persecution will be biased. Despite this potential bias, as we shall see, we can still identify significant negative implications of persecution on the lives of SHARE respondents.

The SHARELIFE questionnaire contains an entire section focused specifically on persecution. Two main questions in this section address the general experience of persecution and dispossession in the following manner:

- There are times, in which people are persecuted or discriminated against, for example because of their political beliefs, religion, nationality, ethnicity, sexual orientation or their background. People may also be persecuted or discriminated against because of political beliefs or the religion of their close relatives. Have you ever been the victim of such persecution or discrimination?
- There may be cases when individuals and their families are dispossessed of their property as a result of war or persecution. Were you or your family ever dispossessed of any property as a result of war or persecution?

These main questions for those who report persecution or dispossession are followed up by detailed questions on the form and consequences of persecution as well as timing and form of dispossession.

The sample we use for the analysis is based on SHARELIFE data for individuals who could be merged with earlier waves of the survey. The initial sample includes 23,981 individuals from 13 countries, of which 22,897 gave valid information to the persecution and dispossession questions. This sample is used for the initial descriptive analysis of persecution data for all countries. The more detailed analysis in Sects. 24.3 and 24.4, which focuses on the Czech Republic, the former German Democratic Republic (GDR, East Germany) and Poland, uses information from SHARELIFE and some matched data from previous waves of SHARE concerning health and life satisfaction outcomes. Since from the German sample we only use data on those who prior to 1989 lived in the East Germany, the sample is much smaller compared to those in the Czech Republic and Poland. Some sample statistics on the three central European populations are given in Table 24.1. In total we have valid information on persecution questions for 1,815 individuals from the Czech Republic, 430 Germans and 1,710 Polish respondents.

Table 24.1 Sample statistics: Czech Republic, East Germany, and Poland

Ever persecuted	Czech Republic		East Germany		Poland	
	Yes	No	Yes	No	Yes	No
Number of observations	166	1,650	39	391	102	1,608
Average age	66.6	63.3	64.1	65.1	67.4	63.0
Female (%)	0.52	0.56	0.42	0.56	0.43	57.3

Source: Authors' calculations using SHARELIFE data

24 Persecution in Central Europe and Its Consequences

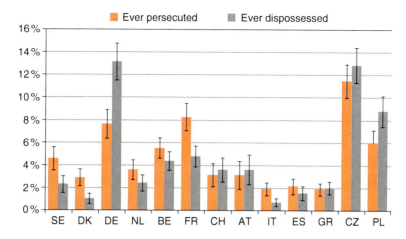

Fig. 24.1 Persecution and dispossession in the SHARELIFE Sample

Figure 24.1 shows the percentage of all respondents that were victims of persecution or whose family was dispossessed. In the Czech Republic, Poland, Germany, France and Belgium more than 5% of SHARE respondents have been persecuted and in the first three countries also dispossessed of their property. In the Czech Republic more than 10% of the current 50+ population have been directly affected by persecution according to the two measures.

The fraction of persecuted persons grows with age: among respondents who are older than 80 years, 18% were persecuted in the Czech Republic, 14% in Poland and 11% in Germany. Among those who prior to 1990 lived in the former East Germany, more than 37% of respondents older than 80 years were dispossessed, compared to only 18% in the former West Germany.

When looking at the data on persecution we must remember that respondents' current countries of residence may not necessarily have been the countries or regimes where they experienced persecution or where they were dispossessed. In fact, in many cases persecution or dispossession may have led or been part of forced relocation. This applies in particular to populations of Germany and Poland, which were subject to significant forced migration as a result of changing borders in the aftermath of World War II. A high number of individuals experiencing persecution or dispossession in other countries were born outside their current country of residence.

In Figs. 24.2 and 24.3 we distinguish between those born within and outside the current borders of countries in which they were interviewed respectively, and the figures confirm the significant relationship between migration and persecution. On average only about 3% of respondents who were born within the borders of their current residence were persecuted and dispossessed. This is in contrast to the average of 17% among those born outside current borders. The figures also explain why we observe instances of persecution in such countries as Denmark or Sweden,

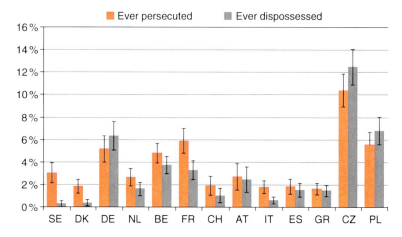

Fig. 24.2 Persecution and dispossession in the SHARELIFE Sample – individuals born within current borders of country of residence

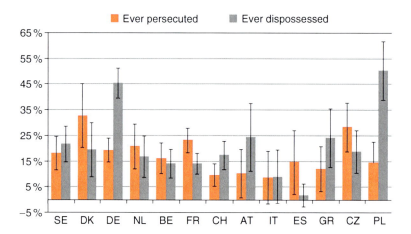

Fig. 24.3 Persecution and dispossession in the SHARELIFE Sample – individuals born outside current borders of country of residence

which are likely to have been hosts to immigrating individuals who escaped persecution.

Interestingly, the results indicate also the different nature of the communist oppression (primarily targeting its own population in order to preserve its power) and of the Nazi regime in Germany and occupation of France and Belgium. While the fraction of the dispossessed in the Czech Republic is relatively independent of the birth place, the high fraction of dispossession among Germans and Poles reflects the resettlement at the end of World War II.

State persecution took various forms, and affected individuals in different countries in many ways. A full account of the instances of persecution, their roots and political background would require a detailed analysis of the European political history of the twentieth century, which is beyond the scope of this chapter. Thus the next section focuses on three populations which, in political terms, have been subject to the same form of state oppression, namely the Czech Republic, the former East Germany and Poland. Each of these countries has been under the communist rule for a significant part of the lives of SHARE respondents in these countries. We show that despite the common roots of the three regimes, their nature, the degree of oppression and the forms of persecution were very different. Our analysis demonstrates that despite the fact that a long time has passed since the individuals experienced direct forms of persecution, and despite the 20 years of democratic experience, the consequences of persecution on individual's welfare, their assessment of life achievements and life satisfaction are still present to this day.

24.5 Persecution and Its Consequences in Central Europe

In this section we focus in more detail on three populations of the SHARE sample with common political history, namely the Czech Republic and Poland, and the "German Länder" which constituted the former German Democratic Republic. For these three samples we examine the differences in the reported form and consequences of persecution. In the analysis we distinguish two types of consequences:

- Direct implications, by which we mean the consequences experienced by the respondents at the time they were subject to persecution. The SHARELIFE interview directly asks for the details described.
- Long-term effects, by which we understand the consequences on the welfare of SHARE respondents, their assessment of their career and general life-satisfaction. The long-term effects are identified either from questions asked independently of the specific persecution questions in the SHARELIFE interview or from previous waves of the survey.

One of the most important features of the data on persecution in the Central European SHARE sample, despite the common history of the communist rule after World War II, are significant differences in terms of the timing and the reported main reason for persecution. In Fig. 24.4 we present the distribution of the (first) experience of dispossession reported by the respondents in the three populations. The differences are striking and they reflect the specific timing of major historical events which affected the lives of respondents.

In the Czech Republic, only 5% of all dispossessions took place during World War II, while over 76% in the period between 1948 and 1956. The timing in Poland and Germany reflects more directly the consequences of the outbreak and duration of World War II, as well as the time immediately following the war. The Polish population suffered from dispossession both by Nazi Germany and the Soviet Union in the period 1939–1943. In the following years it was subjected to forced nationalisation, land dispossession, and forced migration as a result of the

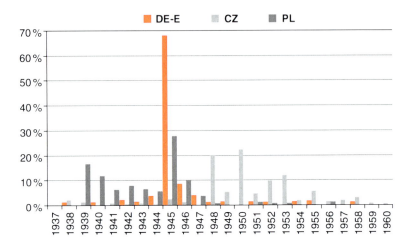

Fig. 24.4 Timing of dispossession in Central Europe

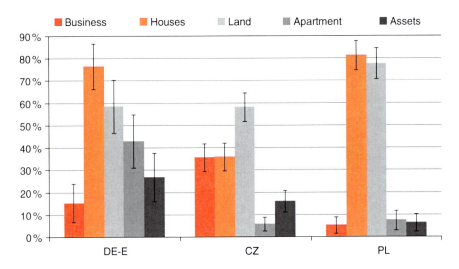

Fig. 24.5 Type of property lost to dispossession in Central Europe

communist take-over from mid-1944 and during the years immediately following the war, and the border changes in 1945. In Poland, 26% of dispossessions occurred in 1939–1940 and 44% in 1944–1947 [those who were born outside the current Polish border lost their property mainly after the war (53%)]. Dispossession among the German population is largely a reflection of the shifting borders and relocation (79%), though East Germans have also been subjected to collectivisation and nationalisation.

Differences between the populations are also evident in the type of property which was lost due to dispossession (Fig. 24.5). This reflects on the one hand the prime focus of nationalisation or collectivisation, and on the other the different nature of main wealth holdings in the middle of the twentieth century in the Czech Republic, East Germany and Poland. Houses and land were the most frequently lost assets in Poland (almost 80% of cases), while land was taken away from 60% of dispossessed East Germans and Czechs. Almost 30% of Czech respondents who suffered dispossession lost their business, a much higher proportion than in the East Germany (14%) and Poland (6%). The type of property lost reflects the timing of dispossession (Fig. 24.4). In Poland and Germany, houses and farms were the main property families lost in forced resettlement during and after World War II. In the Czech Republic, the main type of lost property are farms and farm land during the forced collectivization as well as full nationalization of businesses from 1948 to 1953. The SHARELIFE data also reflect differences in terms of compensation for property lost during the war and communist years. While 47% of Czechs have not been compensated for their lost property, the proportion is much higher among the East German population (54%) and is highest in Poland (65%).

Figure 24.6 presents further interesting differences between the three regimes, and shows information on the reported main reason for persecution, where respondents were prompted to choose from:

- Own political beliefs (1),
- Own religion (2),
- Own ethnicity or nationality (3),
- Own sexual orientation (4),
- Own background (class/property ownership) (5),
- Political beliefs or religion of close relatives (6),
- Or other reasons (7).

Figure 24.6 groups categories (4), (6) and (7) as "other". "Political" corresponds to category (1), and "background" to category (5). Once again there are important

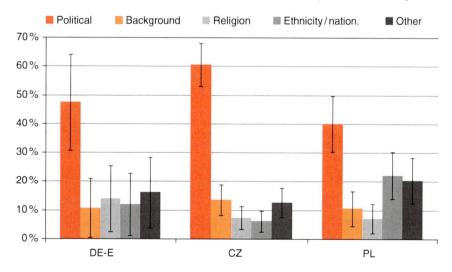

Fig. 24.6 The main reason for persecution in Central Europe

differences in the experiences of the three populations. While political reasons dominate in all countries, they seem to be much more frequent in the Czech Republic (61%) compared to Poland (40%). Ethnicity/nationality was an important reason for persecution in Poland (22%), but least prominent to the current residents of the Czech Republic. Note that ethnicity is also the second most important reason for Poles. Finally, about 10% of respondents in all these ex-communist countries state that their family background, as well as political beliefs and religion of their relatives were also important reasons for persecution. There are important and interesting differences in terms of the main reason for persecution when we compare the former East and West Germany (figures not reported here). While among those who prior to 1989 lived in West Germany and report being subjected to persecution, ethnicity/nationality was the main reason (47%), among East Germans this is the case only for 12% of the persecuted. Only 15% of persecuted West Germans report their political beliefs as the main reason for persecution (compared to 47% in the East). Naturally these differences reflect the post-war political development in the two parts of Germany.

In the communist countries, job-related persecution was one of the most prevalent forms (see Fig. 24.7). This is in particular evident in the data for the Czech Republic. Almost 40% of persecuted people were denied promotions or experienced pay cuts, more than 20% of persecuted persons lost their jobs. 35% of Czechs who experienced persecution were harassed or assigned to tasks of lower responsibility or qualification, and almost 30% experienced difficulties finding a job adequate to their training or qualification. In East Germany these numbers are also high (25% lost their jobs) while in Poland persecution with immediate implications for the job situation was much less frequent. Only about 10% of Poles in the SHARE sample who admit to having been persecuted lost their jobs as a result of persecution,

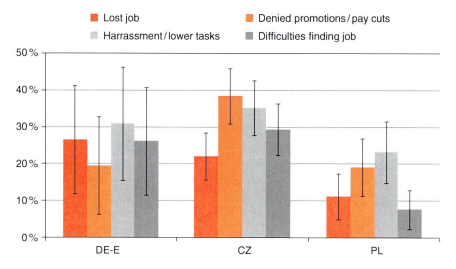

Fig. 24.7 Persecution and direct job implications in Central Europe

and an even lower proportion experienced difficulties finding a job in consequence of persecution.

There is no doubt that an important part of the SHARE sample who lived in the former communist regimes have been subjected to significant stress and discrimination due to their political beliefs, religion or ethnicity. They and their families suffered financial consequences in terms of lost wealth, and in terms of their career development. Although a lot of time has passed since the time when the respondents were subjected to persecution, and prior to the SHARELIFE interview they have lived for almost 20 years under free, democratic regimes, the SHARE data show that the history of persecution does not remain unmarked in terms of their current health and welfare. We turn to these long-term effects of persecution in Sect. 24.6 below.

24.6 Long-term Effects of Persecution

Two types of broad welfare measures have been selected for analysis of long term effects of persecution in Central Europe – six different measures of health, and four different measures of life satisfaction. It is clear that persecution has had detrimental effects on individuals' lives. For example, in the SHARELIFE interview, those persecuted are much more likely to report having experienced periods of stress, poor health, financial hardship or hunger. Although significant amount of time has passed since the periods when individuals were persecuted, as we shall see below the experience of persecution matters for their current situation.

Figures 24.8 and 24.9 report odds ratios for the likelihood of experiencing several forms of poor health (Fig. 24.8) and life satisfaction (Fig. 24.9). In all

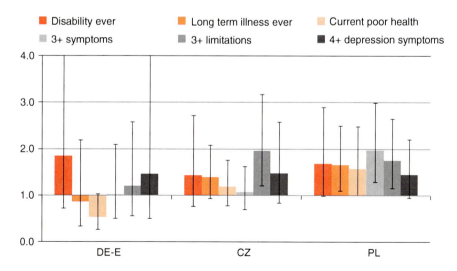

Fig. 24.8 Persecution and long-term effects on health in Central Europe

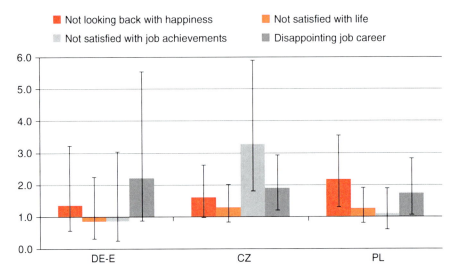

Fig. 24.9 Persecution and long-term effects on life satisfaction in Central Europe

cases we control for age and gender of the respondents, as well as for a number of early life circumstances. The latter are based on several variables from in the SHARELIFE interview including childhood housing quality, relative performance at school, controls for the profession of the main bread winner and number of books in the household at childhood. The black bars correspond to 90% confidence intervals for the odds ratios.

We analyse three measures of health reported in the SHARELIFE interview: current status of general health (reported "fair" or "poor" health), and declarations of experiencing disability or long-term illness. The other three measures are taken from the baseline interview in wave 2, and we examine the odds of having three or more symptoms of poor health, three or more limitations in activity of daily living and declaring four or more symptoms of depression. The experience of persecution in Poland is reflected in odds ratios in excess of one for all measures of health considered, and in almost all of these the effects of persecution are statistically significant. The effects are less significant in the Czech Republic but also strongly above one in four out of six cases. Only in the DE-E sample the effects do not seem to be significant, which partly reflects the much lower sample size of East German respondents.

The experience of persecution does not seem to have strong effects on overall life satisfaction, or the likelihood of looking back at life with happiness (Fig. 24.9), though in the latter case in Czech Republic and Poland there is some evidence on the difference between those who experienced persecution and those who did not. Only in the Polish case this is significant at 90%, and persecuted individuals are twice as likely to report that they do not look back at life with happiness compared to those who did not suffer persecution. It seems that although persecuted individuals have gone through significant periods of stress and various extreme situations, their assessment of life overall does not differ that much from others. Interestingly the SHARELIFE data also suggest that they are more likely to report a specific period of happiness. Such attitudes are also reported in studies on the Holocaust survivors and their children (see Krell et al. 1997; Shmotkin and Litwin 2009). However, when we look at satisfaction with professional life, the consequences of persecution are still very strong, in particular in the Czech Republic, where as we saw in Fig. 24.7, the direct implications on professional life were most severe. The Czech respondents who experienced persecution are three times as likely to be dissatisfied with their job achievements, and almost twice as likely to report dissatisfaction with their careers in comparison to those who did not experience persecution. In Poland persecution increases the likelihood of reporting disappointing job careers by about 80%, and while the effects are not significant in the former German Democratic Republic they are in similar range.

24.7 The Role of Distant Past in SHARE Data

SHARELIFE data provide a unique source of information on the effect of often remote historical events and circumstances on the lives of European populations. The data identify instances of persecution and property dispossession among those currently aged 50+, who have themselves experienced the tragic side of the European twentieth century history.

In this chapter we have presented the information collected in the SHARELIFE interview on often dramatic experiences of European populations in the twentieth century. The data recorded reflect experiences in World War II and in its immediate aftermath, dramatic situations of populations who lived under communist regimes, as well as instances of persecution related to the period of de-colonisation, and other national political distress. A high proportion of those who report having been persecuted, in particular in West European countries, was born outside the current borders of these countries, and persecution is likely to have been the key motivation behind their migration. In the entire SHARELIFE sample almost 5% responded that they have experienced being persecuted, with the highest proportion in the Czech Republic (12%) and lowest in the south of Europe (about 2%). Persecution in the twentieth century Europe often led to the loss of lives or had extremely severe effects on health, which may have led to a significant shortening of people's lives. It also led to forced migration or to departure of native populations to escape the oppressive nature of European regimes. Any analysis on those who survived and remained in their countries can thus uncover only a "lower bound" of the detrimental effects of persecution on people's lives.

Our focus in the analysis has been on three populations who were heavily affected by both World War II and subsequently the long period of communism, namely the Czech Republic, the German population who lived in the German Democratic Republic, and Poland. The common past makes the three a particularly interesting case for comparison, both in terms of extent and form of oppression. While in all three cases political reasons were most frequently given as the main reason for persecution, oppression for political reasons was more frequent in the Czech Republic than in Poland or the German Democratic Republic. In the latter two, ethnicity or nationality was on the other hand a more frequent main reason for persecution than in the Czech Republic. In the Czech Republic persecution had more significant effects on the position at work, with greater frequency of harassment and denied promotions.

Although the reported instances of persecution and dispossession in the majority of cases took place a long time ago, the consequences of these situations are felt by the respondents to this day. These effects are evident in the case of long-term health effects and are reflected in lower satisfaction with life and professional career. Interestingly, the country where the effects of persecution on satisfaction with career are strongest, the Czech Republic, is also the country where persecution had most frequent immediate implications on the position on the labour market. The Polish persecuted population seems to have suffered most in terms of their health state as a consequence of persecution, with the persecuted individuals almost twice as likely to report problems in several health dimensions.

Even though our analysis treats instances of state oppression from the distant past, there are lessons from it which have implications for the policy of the European welfare state in twenty-first century. We could divide them into two main areas:

- Significant periods of stress and oppression have serious detrimental long-term effects on health; this places particular responsibility on democratically elected governments with respect to the compensation towards and care of those who experienced persecution in the past. This regards both those who experienced persecution in their home countries in Europe, as well as those who emigrate as a result of persecution and find their home in the EU.
- Although the nature of harassment at work in communist regimes was different to the forms it might take nowadays, it seems that there is a strong link between the experience of harassment and job satisfaction. Appropriate anti-harassment legislation may thus contribute to higher satisfaction with work and as a result perhaps contribute to longer working lives.

The initial analysis using persecution information collected in SHARELIFE interviews presented in this chapter is only the first attempt to uncover the consequences of dramatic events from respondents' life histories on their welfare. In future work we will investigate the effects on health and wellbeing in more detail, on family and social circumstances, as well as on economic implications of persecution, such as employment histories, incomes and wealth accumulation.

References

Beebe, G. W. (1975). Follow-up studies of World War II and Korean War prisoners. Morbidity, disability and maladjustments. *American Journal of Epidemiology, 101*(5), 400–422.

Danieli, Y. (Ed.). (1998). *International handbook of multigenerational legacies of trauma*. New York: Plenum Press.

Krell, R., Sherman, M. I., & Wiesel, E. (Eds.). (1997). *Medical and psychological effects of concentration camps on holocaust survivors.* Jerusalem: Institute on the Holocaust and Genocide.

Rummel, R. J. (1994). Power, genocide and mass murder. *Journal of Peace Research, 31*(1), 1–10.

Shmotkin, D., & Litwin, H. (2009). *Cumulative adversity and depressive symptoms among older adults in Israel: The differential roles of self-oriented versus other-oriented events of potential trauma. Social psychiatry and psychiatric epidemiology.* New York: Springer.